Air Pollution Control Equipment Selection Guide

Air Pollution Control Equipment Selection Guide

Third Edition

Kenneth C. Schifftner

CRC Press
Taylor & Francis Group
Boca Raton London New York

CRC Press is an imprint of the
Taylor & Francis Group, an **informa** business

Third edition published 2021
by CRC Press
6000 Broken Sound Parkway NW, Suite 300, Boca Raton, FL 33487-2742

and by CRC Press
2 Park Square, Milton Park, Abingdon, Oxon, OX14 4RN

© 2021 Taylor & Francis Group, LLC

First edition published by CRC Press 2002

CRC Press is an imprint of Taylor & Francis Group, LLC

ISBN: 978-0-367-86091-2 (hbk)
ISBN: 978-1-032-00010-7 (pbk)
ISBN: 978-1-003-03257-1 (ebk)

Typeset in Palatino
by KnowledgeWorks Global Ltd.

Contents

Preface

One of the rewards of writing a technical book is to see the finished book on the reference shelf or in the library of colleagues. It is even better to see smudges on the pages, dog-ears, or page markers, thus indicating that the book had been used...perhaps multiple times. The Second Edition has found its way onto the shelves of many air pollution control professionals and its content has been put to use. For that, this writer is exceedingly grateful. In addition, the book has been used as an educational tool for those interested in the control of air pollutants. Favorable comments include that it is "easy to read". That trait has been continued in this edition.

For the pollution control professional, the owner or reader of a technical book such as this, the information contained therein can at the very least assist them in the tasks associated with their job. "Educational" is fine but "supportive" is better. A book such as this should be a resource. A task may be the search for new air pollution control equipment for a new process or project. The goal may be replacement of an existing worn-out device. Or maybe they need to upgrade existing hardware to meet a new code. Perhaps the task is to reduce operating costs. Educating the operators of the equipment may be on the "to-do" list. Or maybe they just want to know what makes the hardware "tick". Any or all of these tasks may be an inherent part of the reader's job or interest. The writer's goal is to produce the content that helps the reader achieve success in meeting those goals.

With technology, times change. Additional information is learned through experience. New concepts are developed and applied. Improved operating techniques are explored and tested. And, yes, lessons are learned from failures. One must keep up with the technology in order to achieve success. For that reason and others, technical books are periodically updated and, fortunately, technical publishers such as Taylor & Francis are willing to support the reader by publishing new and up-to-date content. Thus, this Third Edition was produced.

When information is published for a technical audience, folks invariably ask questions. Many of those questions involve the transition from general knowledge to specifics. For example, "Can you go into greater depth about _____?" Or "I now have the hardware, but how do I integrate that equipment with the process?" Or "How can I reduce the operating costs? Or "I don't think what I am operating is functioning at its peak performance. What can be done?" As a writer, the task is to listen to those questions and supply supporting content that helps provide answers. It is beyond the scope of any such book to supply specific answers since the range of problems to be solved is so broad. There are thousands of processes to which air pollution control equipment has been applied. What can be done is to provide information regarding certain aspects of the application of air pollution control equipment that can be applied to the answering of those questions.

In this edition, a concerted effort has been made to provide insights for answering those types of questions and more. In the Second Edition, a chapter was added about Diagnostics and Testing. In this edition, a comprehensive group of chapters was added regarding *Optimization* to take that information a few steps further. The latter should be particularly helpful to those who need to integrate the gas cleaning equipment with a process (current or future). This new chapter group contains numerous suggested methods for the optimization of air pollution control equipment as applied to common processes. It is hoped that the investment in this book will easily be recovered when some or all of the methods are applied. And the air we breathe will be a bit cleaner. And the cost of running the equipment will be a bit reduced.

For Optimization, success is achieved by maximizing the things that help and minimizing the things that hinder. Obviously, knowing what can help and knowing what can hinder will determine that success. The new Optimization chapters are designed to provide the reader some basic (and hopefully very useful) information that can lead to the successful operation of the hardware defined and described in other sections of the book. Each chapter is designed therefore to integrate with the other chapters of the book. It is expected that the reader will, upon reading an interesting portion of the individual Optimization chapter, go back to the specific chapter dealing with that type of hardware. Much of the new chapter content comes from experience in over 50 years of applying and operating air pollution control equipment. As with previous editions, the knowledge of fellow air pollution control professionals was incorporated to support the information provided. The author would like to thank the following colleagues who provided valuable content for this book. They are Wayne Hartshorn, Bob Taylor, Deny and Michael Claffey, Joe Mayo, Joseph Colannino, Dan Banks, Dan Dickeson, Jerry Childress, Steve Jaasund, and Dr. Robert Richardson. Thank you all! Undoubtedly, there are "nuggets" of information contained herein that will help secure a place of the Third Edition on the reference shelves of air pollution control professionals or persons interested in the technology used to control air pollution.

Author

Kenneth C. Schifftner has more than 40 years of experience in the area of air pollution control. He has been involved with more than 1000 successful gas-cleaning projects. He earned a BS in mechanical engineering at the New Jersey Institute of Technology. He has authored more than 50 technical articles on gas-cleaning technology and coauthored two books. Schifftner has been an instructor for numerous courses sponsored by the Environmental Protection Agency (EPA), provided academic and corporate technical training seminars, served as an expert witness regarding air pollution control technology, and functioned as a consultant to small and large firms interested in solving air pollution problems. Schifftner has also received four US and foreign patents to date on novel mass transfer devices, which are used worldwide. Schifftner is a former chairman of the Environmental Control Division of the American Society of Mechanical Engineers (ASME). He is an active member of ASME, the Semiconductor Safety Association, and Technical Association for the Pulp and Paper Industry (TAPPI).

1

Air Pollution Control 101

One of the favored features of the second edition of the *Air Pollution Control Equipment Selection Guide* was this introductory chapter. The intent is to provide an overview of the basics of air pollution control and introduce newcomers to some of the terminology that will appear in subsequent chapters. This chapter has been used by some instructors to supplement additional studies particularly in application engineering. For this reason, and others, this chapter with some updates has been included in this 3rd edition.

Having spent more than 40 years in the air pollution control industry, I am still amazed how the basics of air pollution control are misunderstood by so many people.

Our newspapers have numerous articles regarding the need to control toxic or carcinogenic substances, but rarely do you see an article explaining *how* it is done. In this chapter, we will explore the basics of air pollution control and how the devices work and, in doing so, introduce some of the terminology used in the industry.

1.1 It Is Separation Technology

Air pollution control can be generally described as a *separation* technology. The pollutants, whether they are gaseous, aerosol, or solid particulate, are *separated* from a carrier gas, which is usually air. We separate these substances because, if we do not, these pollutants may adversely affect our health and that of the environment. Of primary importance is the effect of the pollutants on our respiratory system, where the impact is most noticeable.

Gaseous pollutants are compounds that exist as gas at normal environmental conditions. Usually, *normal* is defined as ambient conditions. These gases may have, just moments before the release, been in a liquid or even solid form. For the purposes of the air pollution device, however, the state they are in just prior to entering the control device is what is most important.

Aerosols are finely divided solid or liquid particles that are typically under 0.5 μm diameter. They often result from the sudden cooling (condensation) of a gaseous pollutant through partial combustion or through a catalytic effect in the gas phase. In the latter condition, a pollutant in the gas phase may combine to form an aerosol in the presence of, for example, a metal co-pollutant.

Acid aerosols, such as SO_3, can form in the presence of vanadium particulate that may be evolved through the combustion of oil-containing vanadium compounds. Solid metals in a furnace can sublime (change phase from solid directly to gaseous) in the heat of an incinerator, then cool sufficiently to form a finely divided aerosol.

Solid particulate can be evolved through combustion or through common processing operations such as grinding, roasting, drying, calcining, coating, or metallizing.

Whatever the state of the pollutant is, the function of the air pollution control device is to separate that pollutant from the carrier gas so that our respiratory system does not have to.

Our respiratory system is our natural separation system. Figure 1.1 depicts major portions of the human respiratory system. Large particles are removed

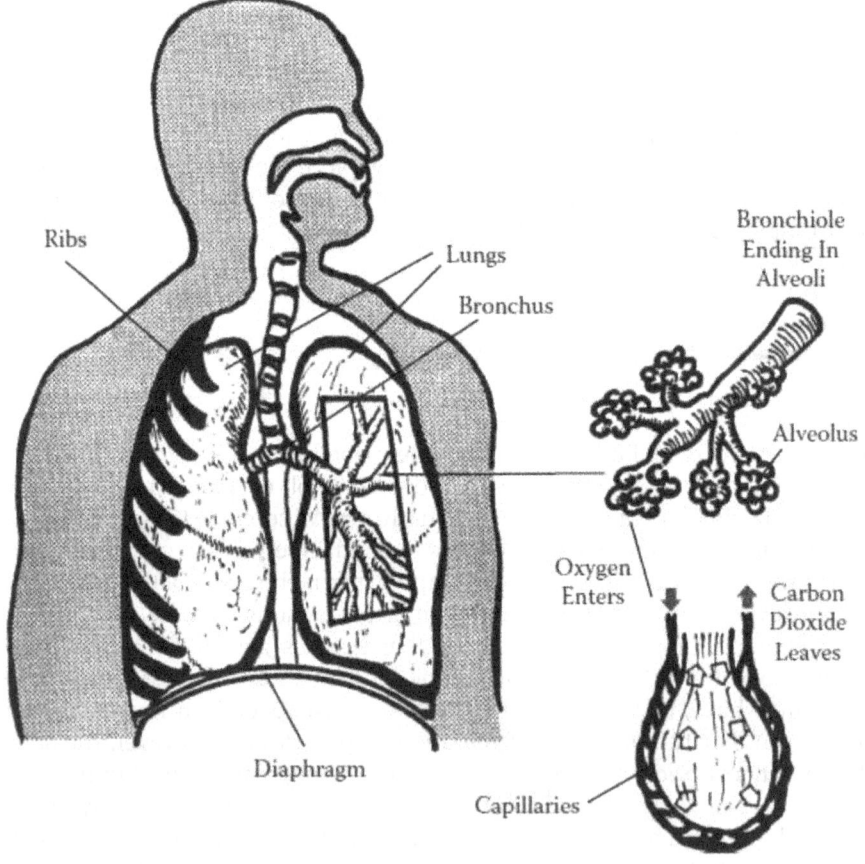

FIGURE 1.1
Respiratory system diagram. (From Marshall, James, *The Air We Live In*, Coward, McCann, and Geoghegan, New York, 1968.)

in the larger openings of the upper respiratory area, smaller particles are removed in the more restricted bronchial area, and the tiniest particles are (hopefully) removed in the tiny alveolar sacs of the lungs.

Air pollution control truly mimics Mother Nature in its separation function. In general, low-energy input wet-type (those using water as the scrubbing medium) gas-cleaning devices remove large particles, higher-energy devices remove smaller particles, and even higher energy (or special technology) devices remove the finest pollutants. In order of decreasing pollutant size, it goes like this:

Mother Nature	Man-Made Devices
Upper respiratory	Low-energy input
Bronchial	Moderate-energy input
Alveolar	High energy or special technology

The larger the particle, or liquid droplet for that matter, the easier it is to separate from the carrier gas.

These characteristics were codified into a helpful chart known as the Frank chart, shown in Figure 1.2. It was named after its creator, an engineer at American Air Filter. This chart shows common particulate sizes and general types of collection mechanisms and devices used for their control. The pollutants are grouped by their settling characteristics. Larger particles (above about 2 μm aerodynamic diameter) generally follow Stokes' law regarding settling velocities. Below about 2 μm, a correction factor (Cunningham's correction factor) is needed to adjust Stokes for the longer settling times for these particles.

In general, particles greater than 20 μm aerodynamic diameter can be controlled using low-energy wet-type devices. Subsequent chapters will explore these devices in detail. These are knockout chambers (traps or settling chambers), cyclone collectors, mechanically aided wet scrubbers, eductors, fluidized-bed scrubbers, spray scrubbers, impactor scrubbers, and Venturi scrubbers (low energy).

For particles 5 μm aerodynamic diameter and above, the Venturi scrubbers (moderate energy) are the most common devices in use. Some vendors have improved the performance of low-energy devices sufficiently to span the gap between those capable of removing 20+ and 5+ μm pollutants. Some mechanically aided wet scrubbers also bridge this gap at higher energy input. For lower concentrations of particles in this size range, enhanced scrubbers such as air/steam atomized spray scrubbers and some proprietary designs are used.

For particles below 5 μm aerodynamic diameter, higher-energy input devices are typically used, or techniques are applied to enlarge these particles to make them easier to capture. Such designs are Venturi scrubbers (high energy), air/steam atomized spray scrubbers, condensation scrubbers, and

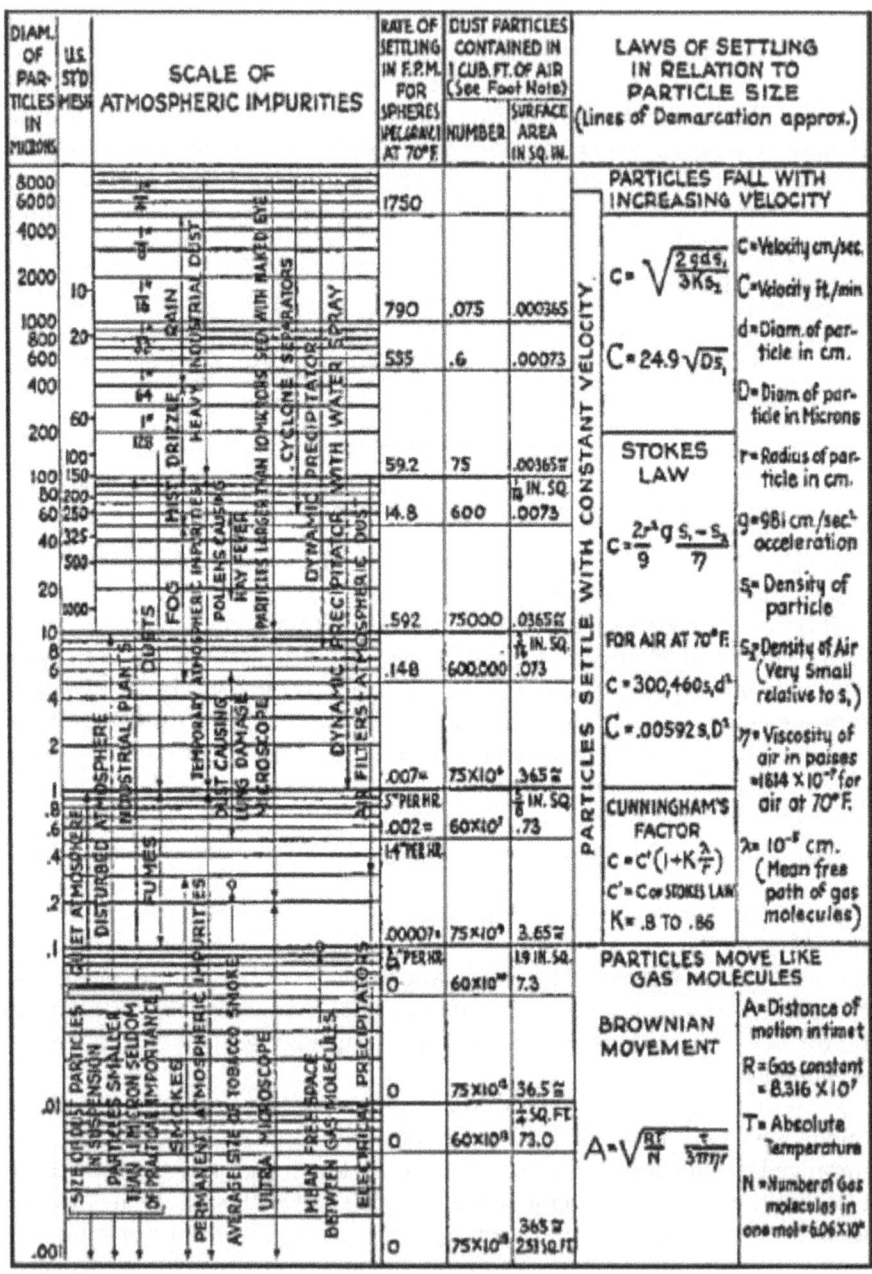

FIGURE 1.2
The "Frank" chart (American Air Filter).

combination devices. If the inlet loading (concentration) is less than approximately 1–2 grs/dscf (grains per dry standard cubic foot), electrostatic forces can be sometimes applied. These include wet electrostatic precipitators and electrostatic scrubbers.

For dry-type separation devices, such as fabric filter collectors (baghouses) and electrostatic precipitators, the energy input is constant regardless of the particle size. Even among these designs, however, increases in energy input yield increases in the collection of finer pollutants. Baghouses are often precoated with a fine material to reduce the permeability of the collecting filter cake and improve fine particulate capture. This cake adds to the pressure drop, which mandates, in turn, an increase in energy input. Precipitators are often increased in field size to remove finer particulate thereby requiring greater power input. These dry devices, in general, use less total power input than equivalent wet devices when removing particulate.

1.2 Wet Collection of Particulate

Wet scrubbers exhibit an increase in total energy input as the target particle size decreases because of the capture technique used.

How is particulate removed in a wet scrubber?

Studies of particle settling rates and motion kinetics have shown that particles greater than approximately 2–5 µm behave inertially and smaller particles tend to behave more like gases. For the former, if you could throw a particle like a baseball, it would follow a given trajectory (perhaps curve or slide but generally follow a given path). Particles less than approximately 2 µm diameter tend to be influenced by gas molecules, temperature and density gradients, and other subtle forces and do not follow predictable trajectories. If you could throw one of these particles, it might turn and hit you in the face. These are the "givens" in the wet scrubber design equation.

Nearly all wet separation devices use the same three capture mechanisms:

1. Impaction
2. Interception
3. Diffusion

Basically, wet scrubbers remove particulate by shooting the particulate at target droplets of liquid.

Figure 1.3 shows a target droplet being impacted by a particle. The particle has sufficient inertia to follow a predicted course into the droplet. Once inside the droplet, the combined particle/droplet size is aerodynamically

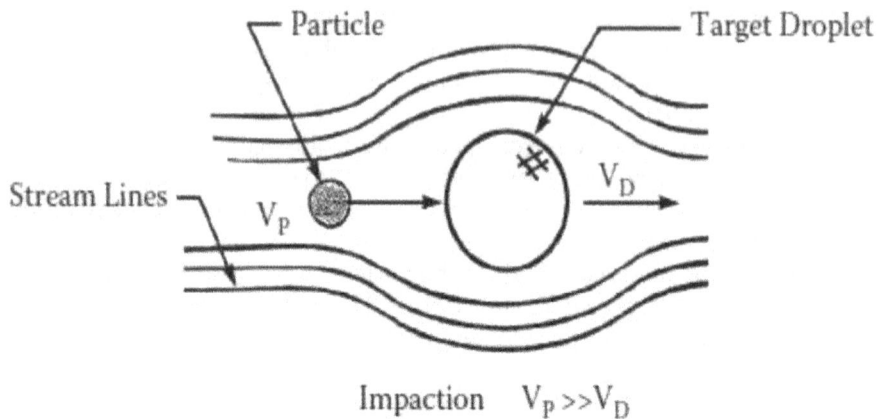

FIGURE 1.3
Impaction (Bionomic Industries, Inc.).

much larger, and therefore the separation task becomes easier. Now simply separate the droplet from the gas stream (more on that later) and the particle(s) are removed.

Figure 1.4 shows a particle, perhaps a bit smaller, moving along the gas streamlines being intercepted at the droplet surface. The particle in this case comes close enough to the droplet surface that it is attracted to that surface and is combined with the droplet. Again, once the particles are intercepted, the bigger droplets are easier to remove.

Figure 1.5 depicts an even tinier particle that is so small that it bounces around in the moving air stream buffeted by water and gas molecules. In this case, the particle *diffuses* over to the droplet and, by chance, is absorbed into

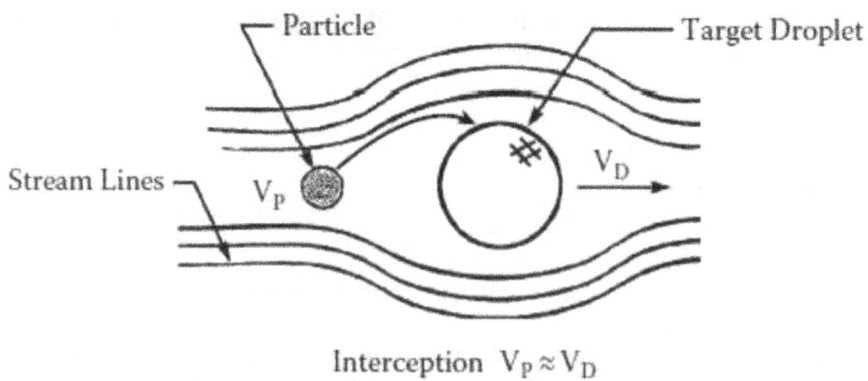

FIGURE 1.4
Interception (Bionomic Industries, Inc.).

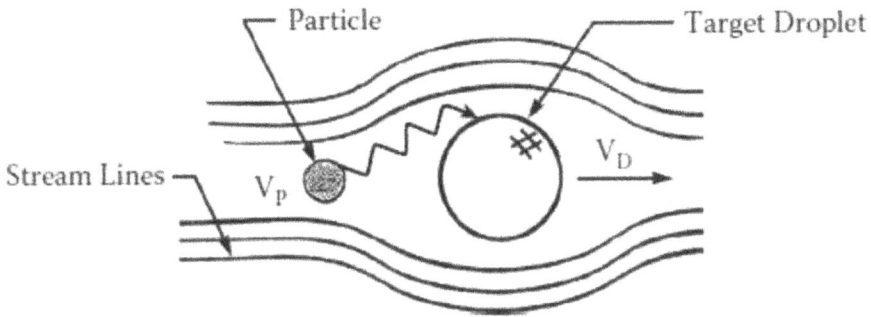

FIGURE 1.5
Diffusion (Bionomic Industries, Inc.).

the droplet. Obviously, to increase the chances of capture by diffusion, one needs to increase the number of droplets per unit volume. This decreases the distance the particle must travel and reduces the chances that it might miss a droplet. Experience has shown that the smaller is the target droplet and the closer the droplet is to an adjacent droplet, the greater is the percentage particulate capture. To make greater quantities of smaller droplets requires increased energy input to shear or form the liquid into tiny target droplets. This is evident in common garden hose spray. The higher the velocity out of the nozzle, the finer the spray.

Once the particulate is into the droplet, Mother Nature tends to help us. Luckily, water droplets generally tend to agglomerate and increase in size upon contact. If we spin, impact, or compress the droplets together, they combine to form even easier-to-remove droplets.

In Figure 1.6, we see a Venturi scrubber (left) connected to a typical cyclonic-type separator. This device separates the droplets using centrifugal force. The centrifugal force pushes the droplets toward the vessel wall, where they form a compressed film, agglomerate, accumulate, and drain out of the air stream by gravity.

Sometimes chevron-type droplet eliminators are used. These place a waveform in the path of the droplet (Figure 1.7). The same process occurs. The droplets build up, drain, and carry their particulate cargo out of the gas stream.

Other forces can also be used to separate fine particulate. If we saturate the gas stream with water vapor and then cool the gas stream, the water vapor will condense on the particulate to form water drops. This same event occurs every day in the form of rainstorms. If it were not for the fact that water vapor condenses on micron and submicron particulate during cleansing rainstorms, we would all suffocate. Condensation scrubbing is the manmade version of the rainstorm.

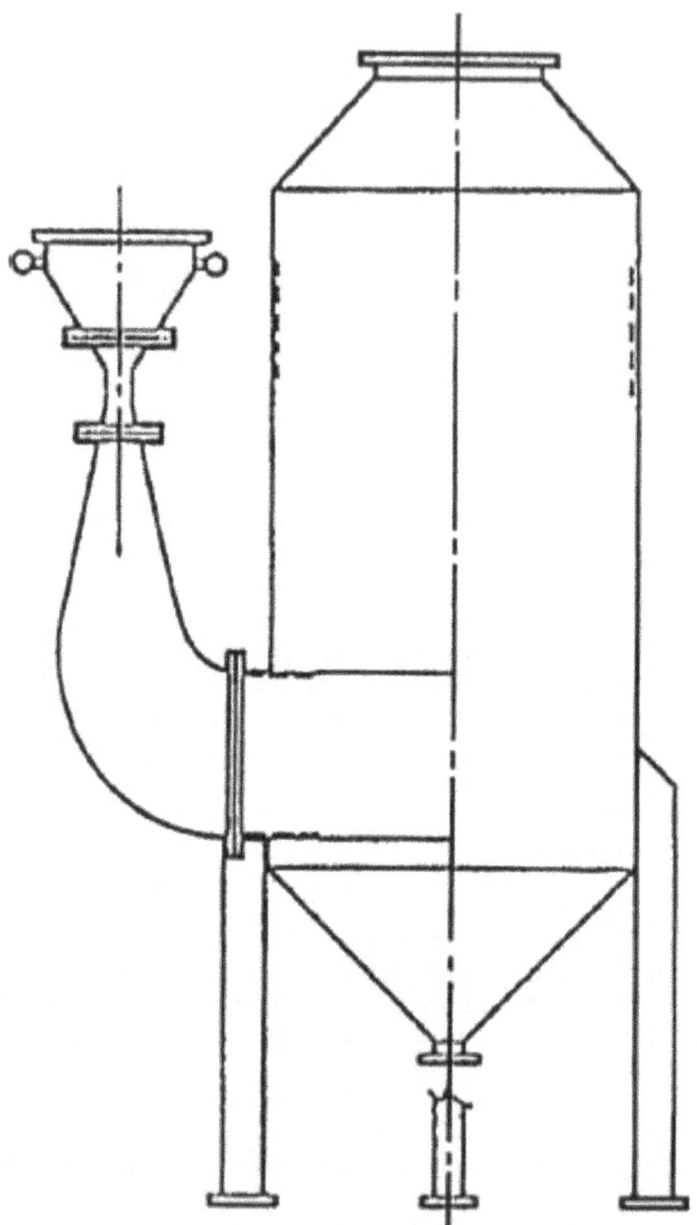

FIGURE 1.6
Venturi scrubber and cyclonic separator.

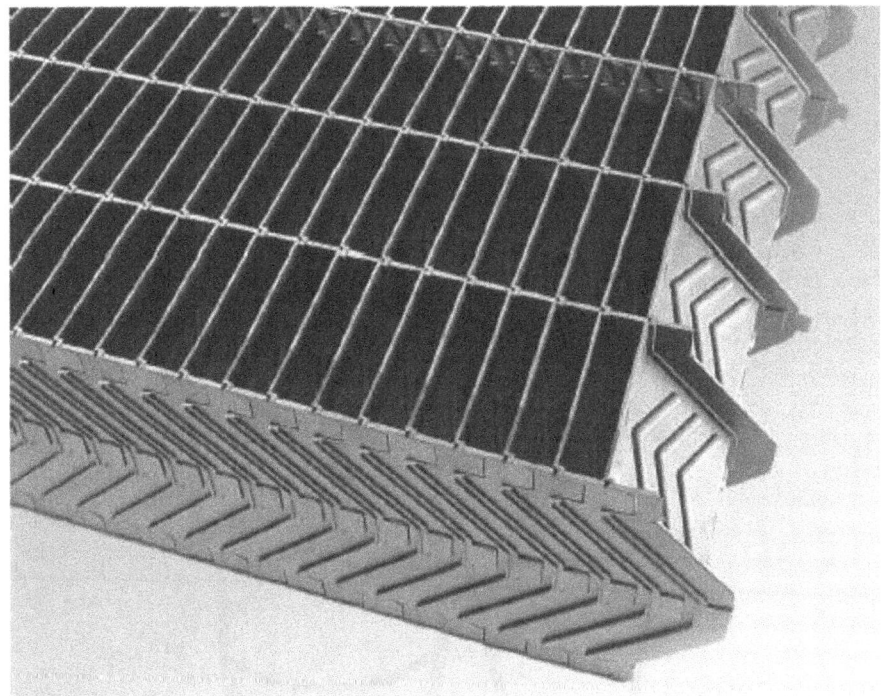

FIGURE 1.7
Chevron droplet eliminator (Munters Corp.).

1.3 Dry Collection

What about collecting the particulate *dry*?

For exceptionally large particles (those greater than approximately 50 μm aerodynamic diameter or about the diameter of a human hair), traps and knockout chambers are used. These basically slow down the gas stream sufficiently so that the particles drop out. These are often seen on the end of lime kilns and mineral calciners as primary separators.

Using the same centrifugal techniques previously mentioned for cyclonic separators, dry cyclone collectors (Figure 1.8) could be used to separate the particulate in a dry form. These devices are commonly used to separate particles more than 5 μm diameter because these particles exhibit the inertia effects mentioned previously. In general, the smaller the cyclone diameter, the smaller the particle that can be removed (because the radius of turn is greater).

To remove more particulate dry, fabric filter collectors or baghouses are used. These devices filter the gas stream through a filter medium, previously removed particulate, or both to remove more particulate. The filter medium is shaken; shaker-type collector, pulsed with air or inert gas, pulse-type

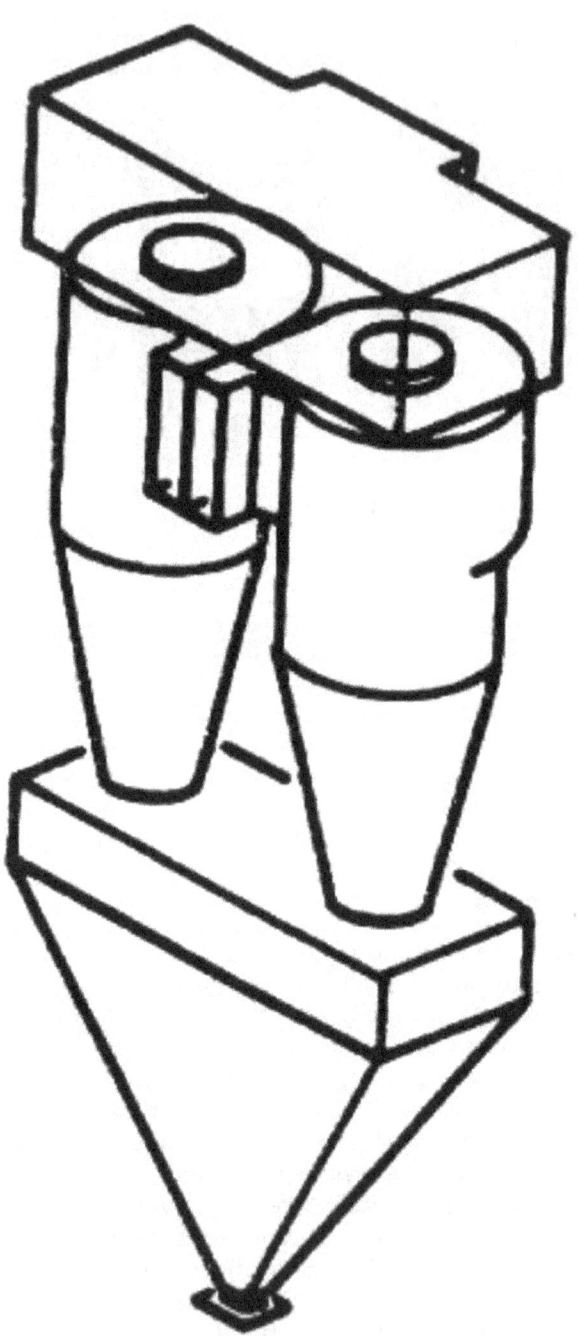

FIGURE 1.8
Cyclone collector (Bionomic Industries, Inc.).

baghouse, or the airflow is reversed to separate the accumulated dust from the filter media, reverse air baghouse.

Subsequent chapters will reveal some of the basics of baghouse selection and design. The general sizing involves selecting the proper filtration medium for the application, the proper cleaning method, and the sizing of the housing velocity or can velocity so that the particulate removed does not entrain back into the gas stream.

For large gas volumes at low inlet concentrations (or loadings) of particulate, dry electrostatic precipitators are used. These units are sized based upon the resistivity of the target dust or particulate (an electrical characteristic) and the particle's ability to migrate to a collecting surface. These parameters determine the electrostatic charge that needs to be applied to charge the particle and the surface area required to collect the particle to a thin enough depth so that it does not insulate the collecting surface and prevent subsequent capture. Subsequent chapters will present the details of precipitator design and selection and how they operate.

1.4 Gas Absorption

What about gases?

In the most basic terms, Mother Nature likes to be in equilibrium. If you blow totally clean air over smelly liquid, some of the smelly gaseous components from the liquid will leave the liquid and seek equilibrium with the gases over the liquid. True equilibrium occurs when the smelly gas ceases to leave the liquid and ceases to return to the liquid.

In the separation of contaminant gases from carrier gases, we help Mother Nature. Figure 1.9 shows a condensing wet scrubbing system.

The processes involved in the separation of contaminant gases from a carrier gas are as follows:

1. Condensation
2. Absorption
3. Adsorption
4. Gas phase destruction (thermal or chemical)

Condensation involves cooling the gas stream sufficiently to condense the contaminant gas. The limit of condensation is the equilibrium condition between the contaminant gas and the carrier gas at the final mixture temperature. For example, a gas stream saturated at 200°F can be condensed to, say, 100°F; however, the resulting outlet gas stream may still contain the amount of contaminant gas that will be at equilibrium with the carrier gas at 100°F.

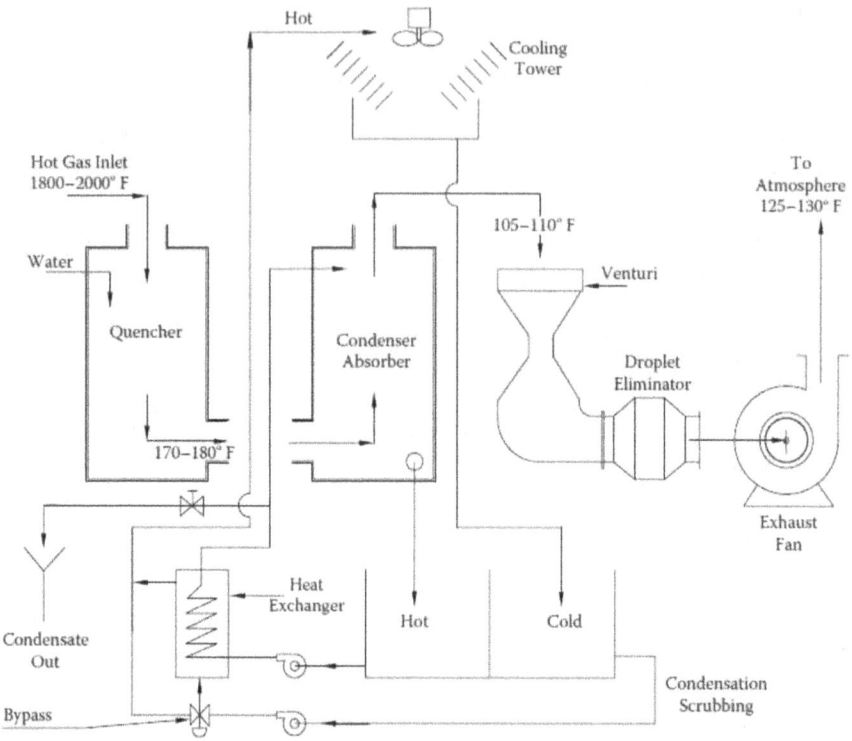

FIGURE 1.9
Condensation scrubbing system components.

Condensation is therefore useful but not always totally effective unless one cools the carrier gas to exceptionally low temperatures.

Absorption is the most common mechanism used in the control of contaminant gases. In general terms, absorption is maximized by:

1. Creating and maintaining the highest liquid surface area to unit gas volume as possible
2. Creating and maintaining a favorable concentration gradient in the scrubbing liquid versus the contaminant gas
3. Doing the above at the lowest energy input

Contaminant gases can enter a liquid stream only at a given number of molecules per unit area. This varies according to the type of contaminant, the type of liquid, the temperature, solubility, and other parameters. In general, however, the greater the surface area of liquid, the greater the amount of gas that can be absorbed and the greater the *rate* at which it can be absorbed.

The leaner (or cleaner) the scrubbing liquid, the greater the transfer of contaminant into the liquid. Gas scrubbers are therefore typically designed to place the cleanest liquid near the cleanest gas (usually at the discharge of the scrubber)

1.4.1 Concept of "Number of Transfer Units" in Absorption

Most gaseous air pollution control processes involve the absorption of the contaminant gas followed by a liquid phase reaction to a salt or oxide that exerts a much lower partial pressure than the raw gas. This process is sometimes called *chemical absorption* or *chemisorption* because the absorption is completed by the subsequent chemical reaction. This action simplifies the design immeasurably because little if any of the absorbed gas wants to strip back into the gas stream. In these quite common cases, the concept of number of transfer units (NTUs) can be used.

The NTUs are defined by the gas absorption process, not by the scrubber design. The NTU concept generally refers to counterflow designs where the liquid moves in the opposite direction to the gas and where the subject gas is absorbed and reacted to form a compound that exhibits little or no vapor pressure at the design conditions, or where the system is very dilute, that is, where the dissolved gas is unreacted but still exhibits little propensity to strip back out of the liquid.

The following explanation of NTUs was contributed by Dan Dickeson of Lantec Products (Agoura Hills, California).

1.4.2 The Transfer Unit Concept in Gas Absorption

Dan Dickeson

Wet scrubbing can be an efficient way to purify air by removing toxic gases that are soluble in water or that can be decomposed by water-based chemical additives.

When water meets air that is polluted with a soluble gas, the water can dissolve only a certain amount of that gas before becoming saturated. Once saturated, it cannot absorb any more of gas. However, the amount of pollutant that can be absorbed by water is not constant; it depends on how polluted the air is. For example, air inside a closed bottle of vinegar contains acetic acid vapor (which is what we smell). When the last drop of vinegar is poured out of the bottle, the smelly air left inside can be purified by pouring some clean water into the bottle and closing it. Acetic acid vapor will dissolve in the water, leaving less and less odor in the air. But as acid is absorbed from the air, the water itself becomes smelly, so it is impossible to remove the entire odor from the air with a single shot of water. This process is illustrated by Figure 1.10.

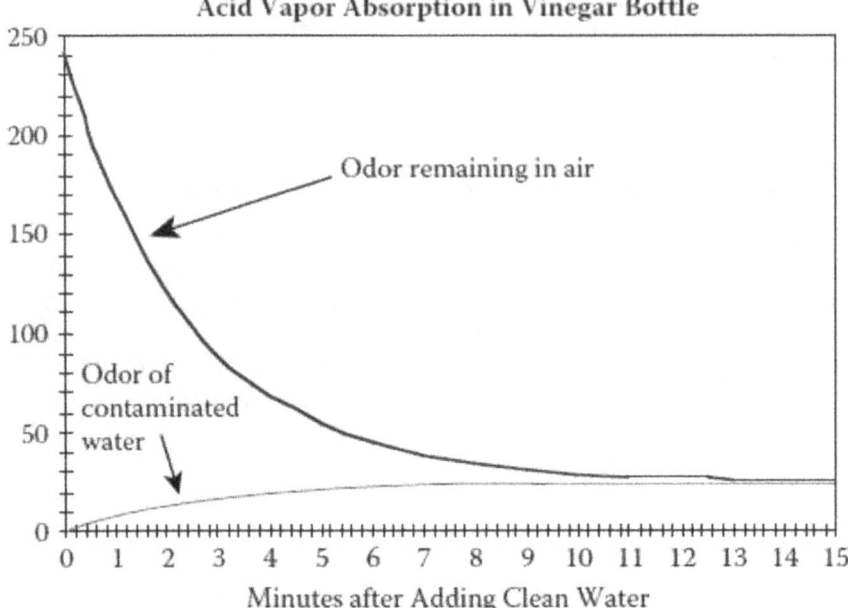

FIGURE 1.10
Equilibrium (Lantec Products, Inc.).

At first, when the water is clean and the air very polluted, acid transfers quickly from the air to the water. But as the amount of acid in the air decreases and the water gets closer to being saturated, the contents of the bottle change more and more gradually. The first 20% of the acid is easy to remove. The last 20% takes much longer to remove. In this example, the last 10% acid is impossible to remove. The closer the two curves get, the more difficult it becomes to absorb additional acid. Chemical engineers define a *transfer unit* as a reduction in pollution by an amount equal to the driving force for absorption (the distance between the curves). This is a useful concept because it turns out that each transfer unit takes the same amount of time to accomplish in a closed system like a bottle or the same amount of *residence time* in a continuous device such as a scrubber.

NTU is a measure of how close a scrubber can come to the saturation limit when purifying polluted air. If neutralizing chemicals are added to eliminate the odor of contaminated water, then there is no limit and the NTU is a measure of how close to zero the pollutant level will come.

Note that the process of odor reduction in the empty vinegar bottle could be sped up considerably by shaking the bottle to bring the air and water into closer contact. In continuous-flow scrubbers, intimate air-water contact is obtained by using packings, froth trays, or spray nozzles to reduce the residence time needed for absorption. The effectiveness of these devices in accelerating absorption is measured by the height of a transfer unit which is the height—or depth—of the contacting section needed to accomplish one transfer unit of purification at a given speed of air flow through it. (Note: NTUs are also described in detail in classic textbooks such as Robert Treybal's textbook *Mass Transfer Operations*, published by McGraw-Hill.)

Because the gas absorption process determines the NTUs and not the device itself, all gas absorbers can be modeled as equivalents. Any absorption problem can be defined in terms of an equivalent of a packed tower, tray tower, fluidized-bed scrubber, spray tower, a mesh pad tower, and so on. There are no miracles that somehow allow a particular design to avoid the realities, the chemistry, of gas absorption. The concept of NTUs makes it easy to compare devices.

The NTUs can be expressed simply as follows:

$$NTU = \ln(\text{concentration IN/concentration OUT})$$

where, ln is the natural log.

The NTUs required are simply the natural log of the ratio of the inlet concentration to the desired outlet concentration.

For example, the NTUs required to reduce an inlet loading of 1500 parts per million by volume (ppmv) hydrochloric acid to 5 ppmv when scrubbing with caustic (low vapor pressure sodium chloride is produced) would be:

$$NTU = \ln(1500 / 5) = \ln(300) = 5.7$$

This means that 5.7 transfer units supplied by any absorber of any design will be required to reduce the hydrochloric acid inlet from 1500 ppmv to 5 ppmv when scrubbing with caustic.

Vendors of gas-cleaning equipment typically perform tests on their designs to determine the NTUs that their design may be able to produce. A tray scrubber vendor may determine, for example, that each of their trays will provide 0.8 transfer units per tray when operated under normal conditions.

To remove the acid in the previous example, we would need:

(5.7 transfer units required)/(0.8 transfer units provided per tray) = 7.12 trays.

A packed tower with inefficient packing might need 2 feet of their packing to provide 1 transfer unit. They would need:

$$5.7 \times 2 = 11.4 \text{ feet of packing.}$$

A packed tower vendor with better packing may need only 1.5 feet of packing per transfer unit, so they would need:

$$5.7 \times 1.5 = 8.55 \text{ feet of packing.}$$

Please note that the removal efficiency of all these systems would be the same. It is also obvious that for a given inlet loading, the lower the required outlet loading, the higher the NTUs required.

If the gas system is not dilute or does not react with the scrubbing solution, the process gets much more complicated.

Adsorption is a separation process where the contaminant gas becomes physically attached to a medium, usually activated carbon, zeolites, or clays. The contaminant gas is physically attached to the adsorbent's surface or onto the pores in that surface or both. Because the pollutant is physically attached, conditions can often be applied that desorb the pollutant from the adsorbent.

Other desorption methods involve the application of inert gas (such as nitrogen) or heat. In Figure 1.11, we see a wheel-shaped accumulator (concentrator) device that is charged with zeolite. The wheel gradually rotates so that one section adsorbs the contaminant, and the other section is thermally desorbed. The contaminant, in this case a hydrocarbon that has some heating value, is thermally oxidized in a separate section and this heat is used to perform the desorption.

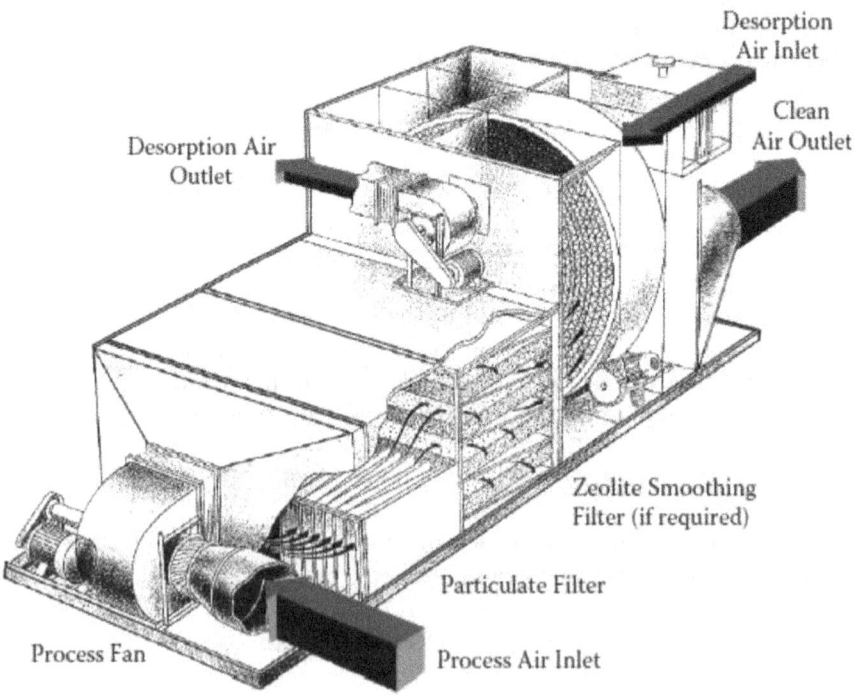

FIGURE 1.11
Rotary concentrator (Munters Corp.).

The design of adsorption systems involves the development of adsorption characteristics for each contaminant compound. These characteristics are graphed, and the result is called an isotherm for that compound. Upon accumulation of the compound into the adsorbent, a point is reached wherein the adsorbent cannot retain any additional gaseous component and breakthrough or bleed through is observed. By regulating the type of adsorbent, its depth, and its time between desorbing (or replacement), the proper removal conditions can be obtained.

Because water vapor can be adsorbed by many of the activated carbon products, water vapor is typically removed prior to an adsorber using carbon. This is accomplished by first cooling the gas stream, then reheating it either using the heat of compression of the fan or by adding supplemental heat.

Gas phase destruction generally occurs in devices called thermal oxidizers. At present, other technologies such as plasma and the application of intense ultraviolet (UV) light are beginning to be explored. In these devices, the chemical bonds of the pollutant are broken through the application of heat, electrical, or light energy.

Thermal oxidizers include direct flame (either open in the form of a flare or enclosed in a refractory or water-lined chamber), catalytic (where a catalyst is used to increase the speed of the bond separation), regenerative (where the heat from the combustion process is used to preheat the incoming gas stream and improve thermal efficiency), and recuperative (where the heat generated is recovered for subsequent use). These devices typically contain a burner that, at least, preheats and initiates the thermal oxidation process and a chamber or housing that contains the products of combustion long enough to allow the desired destruction of the pollutant. In many cases, the pollutant concentration is sufficiently high to allow sustained oxidation without the addition of supplementary fuel.

The residence time in the oxidizer at a minimum temperature has been shown to be an important parameter that controls the ultimate destruction efficiency of the oxidizer. Many regulatory codes require minimum residence times.

For burning solid or mixed wastes, the solid wastes may be first volatilized or converted to carbon, then oxidized in an afterburner. The afterburner becomes the first stage, in effect, of an air pollution control system. This arrangement is common for medical and hazardous waste incinerator systems.

In systems that use UV light, an oxidant (such as hydrogen peroxide) is typically injected into a mixed gas stream followed by the application of intense UV light. The hydroxyl radicals formed by breaking the oxygen/hydrogen bond in the peroxide rather than using free oxygen present in the gas stream attack the pollutant.

1.5 Hybrid Systems

To make life interesting, combinations of two or more of the previously mentioned technologies are not uncommon.

As pollution control regulations have tightened, the need to remove high percentages of each component of a multicomponent pollutant stream has become more important. One control technique may be perfect for one of those stream components; however, it may be totally unsuited for another. For this reason (and others as you will learn in ensuing chapters), hybrid systems combining various technologies are used.

The *order* in which these technologies are used is critical to their success. For example, if ammonia is present in a stream where it might react in the gas phase with another pollutant (say, an acid), the ammonia is usually removed first. This is done so that the ammonia–acid reaction does not form a particulate that would subsequently have to be removed. Another example is the purposeful combustion of sulfurous odorous compounds using a thermal oxidizer, then scrubbing out the sulfur dioxide that is formed using a wet scrubber.

The condensation scrubbing system mentioned previously may include a variety of gas cleaning techniques and even be followed by a wet electrostatic precipitator for fine residual particulate removal.

The combinations used are dictated by the problem to be solved. The problem is broken down into its respective components, suitable technology is selected to control each, then a review is made to minimize or eliminate interferences or redundancies in the control systems. An example of the latter is the use of a wet direct condenser/absorber versus an evaporative cooler on a hot gas cleanup problem. If acid gases and submicron-sized particulate are present and need to be controlled at high efficiency, a wet scrubber can be configured to both subcool the gases and absorb the acid gases. If the acid gas content is minor, an evaporative cooler could be used followed by a baghouse or precipitator. If the acid gas content is somewhere in between and the plant does not have water treatment capability, a spray dryer (dry scrubber) followed by a baghouse or precipitator might be a better choice.

When sizing the actual equipment, the designer must know (or estimate) the gas volumetric and often mass flow rates through the system. For most gas streams, the conditions approximate atmospheric conditions in which the systems operate at relatively low pressures (within a few pounds' gauge of atmospheric pressure) and temperatures of under 2000°F. For others, such as syngas or proprietary process gas streams used in the semiconductor industry, the gas properties may be radically different from those at near atmospheric conditions.

The selection of an air pollution control device starts with an understanding of the gas stream parameters of every point in that specific application. To determine those parameters, the effort may range from source testing of existing emissions points to making guesstimates of the gas flows and characteristics.

Air pollution control equipment is usually sized based upon the actual volume, temperature, and pressure of the gas or gas mixture that must pass through the unit at that stage of the separation process. In the design of some devices that use the kinetic energy of the gas stream (calculated from the mass flow rate), the gas mass flow must also be used. The gas mass flow rate is also used, when coupled with the specific heat of the gas or gas mixture, to determine the stream's heat-carrying capacity and characteristics. The specific heat and molecular weight of the gas stream are used to calculate the psychrometric conditions through the system.

For dry devices such as cyclones and fabric filter collectors, the actual maximum and minimum volume, pressure, and temperature of the gas or gases passing through the device are the most important parameters. For wet devices such as Venturi scrubbers, absorbers, and biofilter-type designs, the saturated gas conditions are most often used. For fluidized devices (wet or dry), the mass flow rate becomes critical since the kinetic energy of the stream is used to provide the desired gas cleaning results.

For most gas-cleaning devices, the design conditions are at or near standard air properties, that is, the gas stream near the ambient atmospheric conditions. Given that costly energy is consumed to pressurize or evacuate gas streams, economics tends to keep most process emissions points near atmospheric conditions (within an atmosphere of pressure). Some processes such as syngas-producing gasifiers, however, must run under considerable vacuum or pressure. In any case, the actual volume, pressure, and temperature of the gas or gas mixture passing through the gas cleaning device must be known.

In the case of wet scrubbers or spray dryer-type systems, the humidity of the gas stream must also be known. The uncontrolled gas stream humidity along with the pressure and temperature determine the saturated gas conditions for a wet scrubber and for a semidry or spray dryer system define the evaporation limits of that device. In a wet scrubber, if it is functioning properly, the gas stream is fully saturated with water vapor as the gas exits the device. In a spray dryer or evaporative cooler, the goal may be to nearly saturate the gas stream while avoiding corrosive condensation. For some dry SO_2 absorption systems, the gas stream is operated near saturation to enhance SO_2 capture but above the acid dewpoint to avoid corrosion. To know how much water can be evaporated, one must know how much water vapor is already entering the device.

Standard air is usually taken as a mixture of about 79% by volume nitrogen, 21% by volume oxygen, 0.04% carbon dioxide, and the rest argon and trace gases. In English units, the density at 1 atmosphere (sea level) and 68°F (20°C) and 50 RH (relative humidity) is about 0.075 lbs/ft³. Gas flows through air pollution control devices are often defined in terms of standard cubic feet per minute (scfm). To complicate things, sometimes the scfm is calculated at 60°F or 70°F; sometimes sdcfm or dscfm is used, indicating that the gas flow is defined totally dry (no water vapor). This term is handy because the dry gas mass flow through a gas cleaning device does not change, though its volume may change given pressure and/or temperature differences. The standard air mixture typically has a specific heat of 0 24 BTU/lb.

For some applications, such as syngas gasification systems, the gas mixture departs radically from that of standard air. Gasification systems typically operate with little or no oxygen and instead have significant quantities of hydrogen, carbon monoxide, and methane. These gases tend to reduce the gas density, which in turn reduces the gas stream kinetic energy. These gases also exhibit higher specific heats, meaning that more heat energy can be carried per unit volume than if the gas stream were standard. For these type applications, one starts with the gas stream analysis (either measured or estimated) and the mixture-specific heat and molecular weight is calculated.

Various programs are available to calculate the gas conditions and various mixture molecular weights and specific heats. Psychrometric tables (the "old-fashioned way") are also available from which the gas parameters can be calculated. As part of the related chapters that follow, some details relating to the technology are discussed.

The foregoing hopefully provided the basics, and some important detail, on how air pollution control equipment operates and some of the theories on which the technology is based. Combining the information contained in this chapter and the more detailed information contained in subsequent chapters, you will be able to carefully select the best air pollution control equipment for your application. "Air Pollution Control 101" is just the start. In the following chapters, we will describe various types of technologies that can be used to control your specific air pollution control problem. You will find that nearly any combination of pollutants can be effectively controlled if the proper control technique is applied. This chapter and the ones that follow should make this selection much easier and provide confidence that your ultimate selection is a wise one.

2

Absorption Devices

2.1 Device Type

Adsorption devices consist of adsorptive media, either static or mobile, in a containing vessel through which the gas and its contaminants are passed. The contaminants are adsorbed onto and into pores in the adsorbing media.

2.2 Typical Applications and Uses

Adsorbers are most commonly used for solvent recovery; control of hydrocarbon emissions from storage tanks, transfer facilities, printing operations; and similar processes where volatile hydrocarbons are present. Activated carbon types are also used to control sulfurous odor, such as that from sewage treatment plants. Special impregnated carbons are used to chemically react with the contaminant once it is adsorbed, thereby extending the carbon life. Where the hydrocarbon has recovery value, absorbers are often used after process vents, evaporators, or distillation columns to polish the emission down to regulatory limits. They are also used on process vents in lieu of thermal oxidizers.

Regenerative adsorbers are generally not used where the contaminant is not economically recoverable, or the desorption process has a low yield. For example, cases where adding steam to desorb the carbon results in an unusable water mixture tend to make adsorption less attractive.

Drum-type units are often attached to process tanks to control hydrocarbon breathing or fill venting losses. The gas flow rates are typically low, and these drum-type units can be applied very economically.

Filter-type units are used in ventilation systems for hospitals, clean rooms, auditoriums, bus stations, loading docks, and other environments where adsorbable hydrocarbons may be present.

2.3 Operating Principles

Gas adsorption is the physical capturing of contaminant gas molecules onto or into the surface of a suitable solid adsorbent, such as activate carbon, zeolite, diatomaceous earth, clays, or other porous media. The gas molecule is physically trapped by the pore openings in the medium and accumulates over time until the medium saturates and can hold no more. In some devices, the medium is desorbed in place through the application of a gas such as nitrogen, or steam, to drive the contaminant from the pore openings of the medium. In others, the medium itself is directed to a device where thermal energy (heat) is applied to desorb and recover the medium.

Adsorption is basically a pore surface and size phenomenon. The size of the gas molecule dictates the pore size of the required adsorbent, and the bulk pore area of the adsorbent per unit volume determines the amount of adsorbent required to control the specific pollutant. Adsorbents exhibit certain physical characteristics with respect to pore size. These characteristics are generally called *macropores* and *micropores* as shown in Figure 2.1. As defined by the word prefixes, *macropores* are large pore openings and *micropores* are small pore openings. In practice, adsorbents exhibit a mixture of both. The volume of adsorbent required is controlled by the contaminant gas rate, and the amount of time allowed before breakthrough is permitted to occur. Breakthrough occurs when the pores are effectively filled with the contaminants or interfering compounds.

The process of activating activated carbon is basically one of opening its pores. The carbon can be acid washed then carefully heated in a reducing atmosphere, or it can be otherwise treated to open the available pores.

Various adsorbents reflect known pore sizes and exhibit specific areas per unit volume. Application engineers have developed *adsorption isotherms* for various pollutants as they relate to specific adsorbent types. In the family of activated carbons, for example, there are dozens of different carbon types (peanut shell based, coconut shell based, mineral carbon based, etc.), each exemplifying specific pore size and area characteristics. The adsorption isotherms are used to predict the rate of capture of that pollutant in the adsorbent and to therefore anticipate breakthrough.

Figure 2.2 shows a typical adsorption isotherm curve. Adsorption tends to follow the lessons learned earlier about NTUs and driving force. The concentration gradient is important in adsorption processes because a large gradient tends to fill pores quickly, thereby reducing the probability of continued adsorption at a high rate. The designer therefore must allow for a sufficient volume of adsorbent, not only for its ultimate capacity prior to breakthrough but also for the concentration gradient that may exist. If the contaminant exists in high concentration, the volume of adsorbent is increased and the speed at which the gas flows through the adsorbent is decreased.

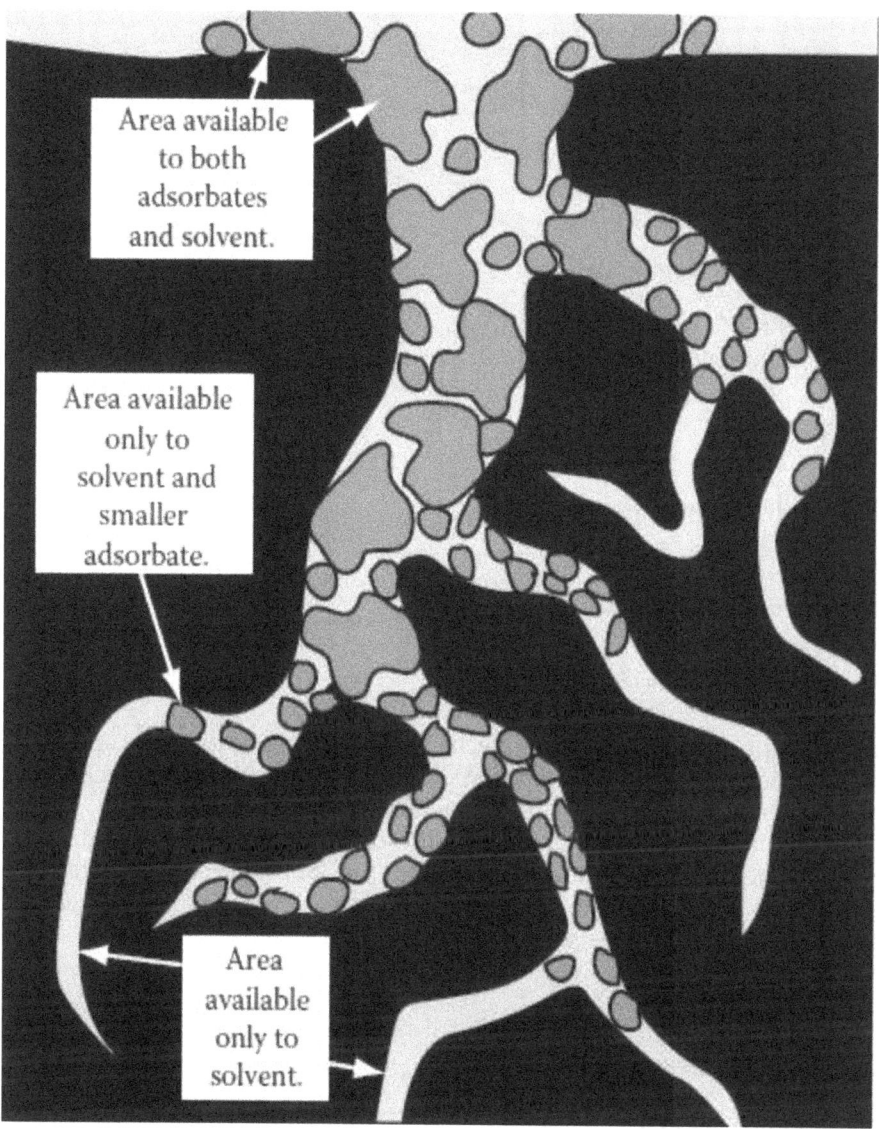

Area available
to both
adsorbates
and solvent.

Area available
only to
solvent and
smaller
adsorbate.

Area
available
only to
solvent.

FIGURE 2.1
Macropores and micropores. (Barnebey Sutcliffe Corp.)

2.4 Primary Mechanisms Used

Although the contaminant gas molecule must be fitted to the available pore size of the adsorbent, the mechanism holding the molecule onto the adsorbent is believed to be van der Waals and other weak attractive forces.

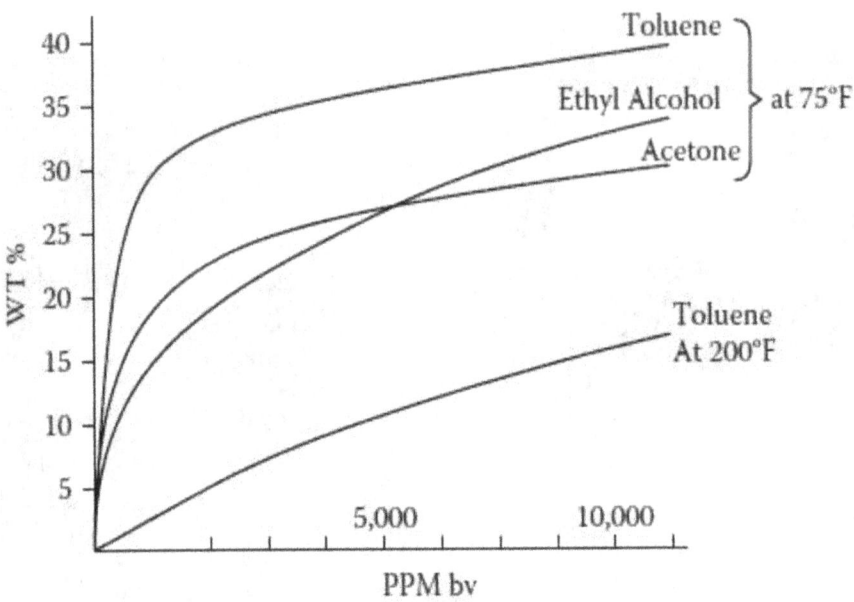

FIGURE 2.2
Adsorption isotherm. (Amcec, Inc.)

The adsorption process is more mechanical than chemical. An exception to the latter is chemically treated adsorbents wherein the pores are precharged with a chemical that reacts with the contaminant upon contact.

Given that the contaminant molecules are mechanically attached, they can often be detached or desorbed through the application of steam, heated gases, inert gases, or other processes that force the contaminant out of the pores. In this manner, the adsorbent can be regenerated and reused to some extent until the useful life of the adsorbent is reached.

2.5 Design Basics

Adsorbers are usually either of the throwaway or regenerative type. The throwaway type involves the use of a fixed bed of adsorbent in a containing vessel. These vessels can be either periodically emptied of the adsorbent or the entire chamber with adsorbent can be exchanged for a new one. The adsorbent is either regenerated remotely or thrown way. In the regenerative type, the adsorbent is regenerated or desorbed in place. This typically involves two chambers that can be isolated. One chamber is actively adsorbing while the other is being desorbed with either steam, hot air, or an inert gas such as nitrogen.

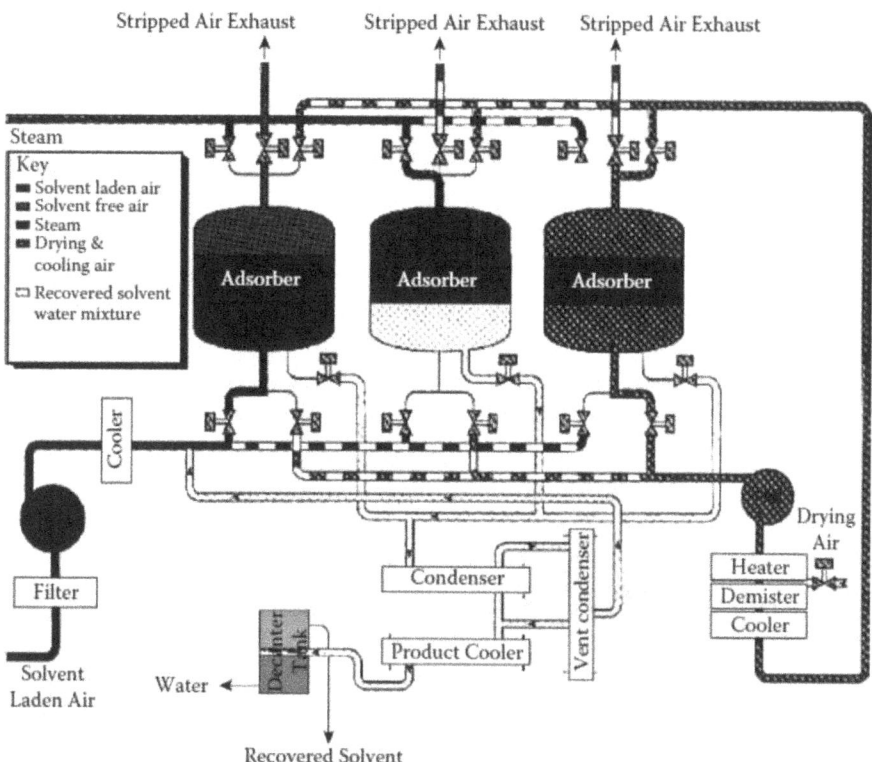

FIGURE 2.3

Regenerative adsorber. (Barnebey Sutcliffe Corp.)

The ancillary equipment includes dampers to swing the contaminant gas stream from one chamber to the other and isolation valves and controls to administer steam to desorb in situ. Some of these designs use an inert gas such as nitrogen for desorption purposes. The desorbed vapors are often condensed and collected or are directed to a thermal oxidizer for destruction. Figure 2.3 shows a multiple chamber adsorber schematic for capture and recovery of solvent-laden air and regeneration in situ using steam.

Sometimes, the designer creates a deep bed of adsorbent and installs it in a modular housing. These are popular for point-of-use volatile organic compound (VOC) control. Equipped with its own fan and pressure-drop monitor, the packaged unit is simple to install and operate. When the adsorbent is consumed (breakthrough occurs), the adsorbent housing can be shipped for regeneration off-site. Figure 2.4 shows a packaged, deep bed-type adsorption unit.

Adsorber gas velocities are usually extremely low to reduce the pressure drop of the system. Because the adsorbent particles are close together, their

FIGURE 2.4
Packaged adsorption unit. (Barnebey Sutcliffe Corp.)

resistance to gas flow is quite high. Gas velocities of 1–3 ft/sec or less are common. The bed depth is dictated by the calculated volume of adsorbent needed to operate before breakthrough based upon the adsorption isotherm(s) for the contaminant(s) to be removed. To avoid channeling of gases, multiple beds are sometimes used. Each bed may be 1–2 ft thick followed by a vapor space to permit gas redistribution. This low gas velocity means that adsorbers are generally large devices.

A throwaway-type (drum) adsorber is shown in Figure 2.5. The adsorbent is precharged in the drum and the drum is designed for off-site regeneration or disposal.

These designs are often used for tank vent emission control for volatile hydrocarbons where the gas flow rate is 50–150 actual cubic feet per minute. Upon achieving breakthrough or scheduled replacement, the canister is removed from service, sealed, and shipped to the supplier for off-site regeneration or replacement.

Unfortunately, water and water vapor can be adsorbed as well on most adsorbents (exception: zeolites). The water vapor becomes, in effect, an unwanted contaminant because it takes away adsorbent area that would be better used to collect the real contaminant. To reduce water's effect on the adsorbent, humid gas streams are sometimes reduced in water vapor content

FIGURE 2.5
Canister-type adsorbers. (Carbtrol Corp.)

by first cooling the gas stream to condense water vapor, then reheating the stream to be well above the water dewpoint. The adsorber housing is then insulated to prevent the water from cooling and reforming a vapor. In low humidity applications, the gas stream is sometimes sent through a bed of gravel or rocks to remove entrained water vapor. Sending the gases through a strong acid scrubber can also dry the gases so that the adsorption process is maximized.

The canister-type systems often include a bed of gravel or a separate water trap canister to reduce the carryover of water to the adsorption canister. Others are band heated to keep the gas humidity below the dewpoint. Sometimes, heated air is bled into the system to reduce the gas moisture content. The most effective method, however, involves cooling the gases to condense water followed by indirect reheat.

If the contaminant gas easily desorbs and can exceed the lower explosive limit (LEL), the adsorber vessel must be designed for explosion-proof operation. The adsorption process is one of methods for concentrating a dilute gaseous stream, so LEL considerations must be considered.

The activated carbon-type adsorbers are generally used in applications of less than 150°F. For higher temperatures, zeolites are often used. Zeolites are mineral-based adsorbents that are less affected by water vapor and temperature. Zeolites have been effectively used in rotating wheel-type

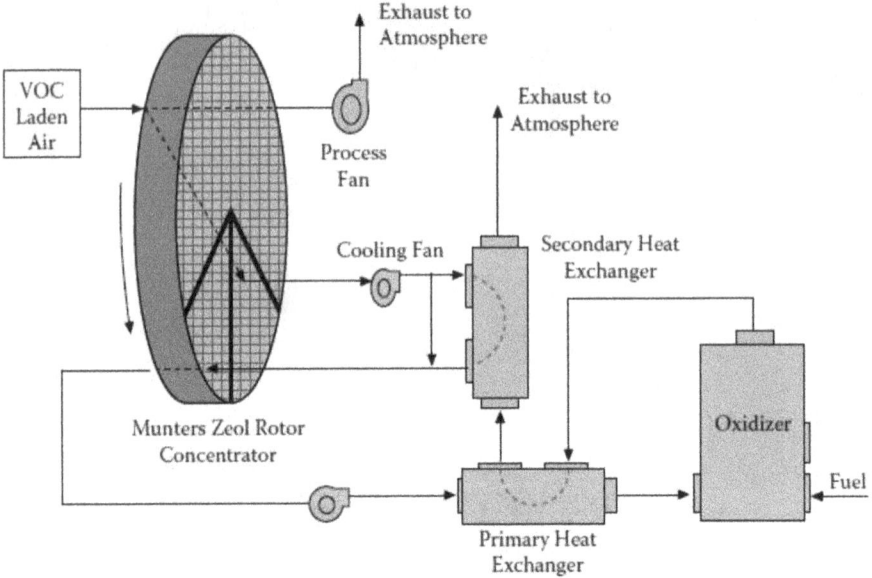

FIGURE 2.6
Zeolite-type adsorption concentrator (Munters Zeol).

devices as shown in Figure 2.6 and as mentioned in Chapter 1. They are used ahead of thermal oxidizers to concentrate the contaminants in a dilute gas stream to a point where they can economically be destroyed thermally. This concentrator-type service reduces the size of the required thermal oxidizer.

Panel-type air filters are also available precharged with activated carbon or another suitable adsorbent. Figure 2.7 shows such a panel filter wherein the finely divided carbon is mixed with the filter medium itself. In other designs, pelletized carbon fills the space between filter medium panels, thereby providing some VOC control. These designs are used in room ventilation systems. The adsorbent, the filter medium, or both can be pretreated with a biocide to kill bacteria that may also be found in the gas stream. Highly specialized filters such as these are used to protect military personnel who handle mobile vehicles such as tanks and personnel carriers from gaseous weaponry and deadly battlefield smoke particulate.

2.6 Operating Suggestions

As previously mentioned, water and water vapor should be removed prior to nonzeolite-type adsorbers. If regenerative type adsorbers are contemplated, the vendor should be consulted regarding the integration of the adsorber

FIGURE 2.7
Panel-type adsorption filter. (Barnebey Sutcliffe Corp.)

into the process and a thorough economic analysis should be performed. On many applications, the use of a regenerative type adsorber can provide significant savings in recovered solvent or chemical.

Except for the rotating wheel-type adsorber, the capacity of any adsorber slowly decreases from the moment of initial operation. As the adsorption gradually moves to the point of breakthrough, the adsorption efficiency stays relatively constant. For this reason, time, or a breakthrough sensor (hydrocarbon analyzer) must be used to determine breakthrough. If batch-type adsorbers are used, one must carefully monitor the time between regeneration or replacement or invest in monitoring equipment that indicates when regeneration or replacement is required.

3

Bioreactors

In air pollution control, *bioremediation* is the use of bacteria and fungi to consume pollutants from an air stream contaminated with organic compounds that are soluble in water or can be solubilized in water. Bioremediation also includes chemical transformation for some inorganic compounds like sulfides and nitrogen oxides.

3.1 Device Types

Biofilters, biotrickling filters, aerobic digesters, and bioscrubber technologies include a wide range of methodologies that are applicable to a diverse spectrum of air-quality applications. Each device type has advantages and limitations. These will be described below.

This chapter also describes recent breakthroughs in bioreactor technology and the combination bioreactors with other air pollution abatement technologies. These breakthroughs include the use of surfactants to expand the range of organic compounds that can be treated with bioreactor methods and breakthroughs that utilize biological organisms in temperature and pH ranges not typically used in biological abatement of air pollution.

After defining the basic vocabularies associated with these methodologies and outlining the fundamental requirements for their success in air-quality applications, this chapter will explain some of the ways these technologies can be combined with conventional air-quality remediation methods to further optimize cost and performance.

3.1.1 Bioremediation

The processes are described as follows.

1. **Biofilter:** An in-ground or above-ground filter bed through which moist air at ambient temperature containing water soluble volatile organic compounds or sulfur-containing compounds like hydrogen sulfide (collectively air pollutants) pass. The bed is made from porous organic or inorganic media that act as a substrate for bacteria or fungi. The air pollutants are removed from the air by transferring

to moisture surrounding the microorganisms which are sequestered to the media bed.

2. **Biotrickling filter:** Similar in design to trickling filters used in the treatment of wastewater but differing in application because these devices treat air laden with pollutants described in the *biofilter* definition above rather than pollutants dissolved in a liquid. Biotrickling filters are typically round vessels with cone-shaped bottoms and no top. A packing bed typically between 5 and 15 ft deep that is made from plastic media with approximately 95% open area is supported above the cone-shaped bottom section of the tank. The packing bed remains moist through recirculated spray of liquid collected in the cone-shaped bottom section of the tank. The sprayed liquid trickles over the packing bed. Air laden with pollutants is passed through the moist packing bed in a countercurrent direction (compared with the flow of the trickled liquid). The air is cleaned as it passes through the packing bed when pollutants are transferred from the air to moist biological slime attached to the packing material. The pollutants are chemically changed by biological organisms in the slime. Solids are periodically removed from the bottom of the cone-shaped section of the tank and dried prior to disposal.

3. **Aerobic digester:** A vessel containing aerated liquid with suspended biological organisms that chemically change pollutants described in the *biofilter* definition above. The liquid can contain free-floating neutral buoyancy porous material that act as substrate for biological organisms. Solids are periodically removed from the bottom of the tank and dried prior to disposal.

4. **Bioscrubber:** Typically, round tanks with lids that contain a packing bed and recirculated liquid system like the design described in *biotrickling filter* above. Air laden with pollutants described in the *biofilter* section above is passed through the moist packing bed in a concurrent or countercurrent direction (compared with the trickling liquid). The pollutants are chemically changed by biological organisms in the slime.

5. **Biofilter technology:** *Biofilters* use biologic colonies that reside on a supporting substrate (biomass) and are selected for their ability to produce enzymes that reduce absorbed organic pollutants to less hazardous or less volatile forms. The biofilter itself is a combination of *adsorber* (the medium on which the bacteria colonize provides an adsorption surface) and *absorber* (the moist biofilm on the medium surface absorbs the contaminants).

   Biofilters are considered by some to be green technology, that is, environmentally friendly. The organic chemical action that occurs within a biofilter is often more complex than that of inorganic chemisorption systems. Biofilters, however, can significantly reduce or eliminate the transportation, storage, handling, and use costs of oxidizing and neutralizing chemicals common to competing wet chemical technology.

3.2 Typical Applications and Uses

Biofilters are often used to control the emissions of water-soluble or condensable hydrocarbons (such as alcohols), phenols, aldehydes such as formaldehyde, odorous mercaptans, organic acids, and similar compounds. They are used to control emissions from aerosol propellant manufacture and filling operations, meat processing and packing processes, pharmaceutical manufacture (fermenter emissions), and fish and other food processing sources.

Candidate pollutants that can be controlled by biofilters, in general, must be water soluble because the biodegradation occurs in the moist biofilm layer supported in the biofilter. Aliphatic hydrocarbons are generally more easily degraded than aromatic hydrocarbons. Halogenated hydrocarbons show an increased resistance to this method as their halogen content increases, although some exceptions exist.

A typical biofilter is shown in cutaway format in Figure 3.1. The basic components consist of a humidification system to produce a saturated gas stream (to the lower left), a substrate to support the biomass, a containing vessel, and some means (such as a fan, upper right) to move the gases through the biofilter.

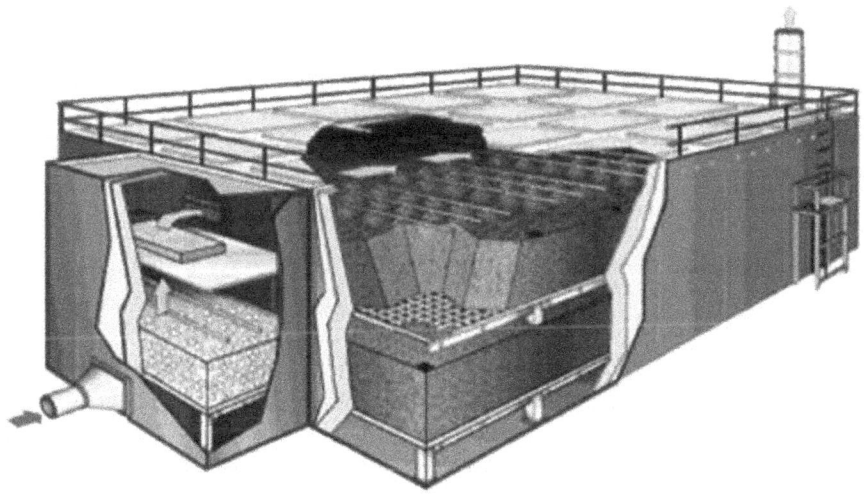

FIGURE 3.1
Biofilter. (Monsanto Enviro-Chem Systems, Inc.)

Biofilters have also been used to control the emissions of propane and hexane from the filling of aerosol cans. These systems can be built into the ground, so the containing vessel becomes the surrounding earth. Buried distribution pipes introduce the contaminated gas beneath the biomass and its support. The gases percolate and diffuse through the biomass layer.

Some meat packing facilities ventilate their meat processing devices (cookers, etc.) into biofilters for odor control. More intense odors are controlled using packed towers, tray towers, fluidized bed scrubbers, and varieties of spray-type devices where oxidizing chemicals are used. These devices can be followed by biofilters, however, wherein the latter act as polishers to remove residual pollutants.

Operating efficiencies of 70%–90% are obtainable with a professionally designed unit with higher efficiencies available if extended residence times are economically feasible. These efficiencies, in the United States at least, are often less than the levels required by the regulatory authorities; therefore, biofilters are not as popular here as in other countries.

To be successful, a biofilter must be used under conditions that are conducive both to the viability of the biofilm and to the absorption of the contaminant. Typical biofilters are operated under 100°F and at 100% relative humidity. They usually operate using a preconditioning spray chamber or scrubber to ensure high humidity. Because the resistance to gas flow through a biofilter is significant, they are often exceptionally large devices. Sometimes, an entire field containing underground distribution pipes is used to provide an adequately large and stable biomass.

Biofilters are used in applications wherein the gas stream does not contain compounds that are toxic to the bacteria and does not contain high particulate loading, and where the gas stream temperature and humidity can be controlled within a range suitable for sustaining the bacteria colonies and where the concentration of pollutants is sufficiently low and sufficiently consistent so that the bacteria colony is not overwhelmed or starved. These conditions vary based on application and bacteria or enzyme selected. Table 3.1 is a list of popular pollutants that are treatable using biologic methods.

3.3 Operating Principles

Bacteria that produce enzymes suitable for the oxidation or reduction of the target pollutant are harnessed to do the work in biofilters. They represent millions of tiny catalytic oxidation sites that in most cases take oxygen in the gas stream and fix it to the pollutant to mineralize it (convert the pollutant to CO_2, water, and innocuous residuals). Some bacteria strains use their enzymes to cleave organic molecules or extract specific elements (such as sulfur), thereby changing the characteristics of the contaminant molecule.

TABLE 3.1

Common Pollutants Recognized as Biodegradable

Acetone	Heptane
Acrylonitrile	Hexane
Anthracene	Isopropyl acetate
Atrazine	Isopropyl alcohol
Benzene	Lindane
Benzoic acid	Methylene chloride
Benzopyrene	Methyl ethyl ketone
Butanol	Methyl methacrylate
Butylcellosolve	Naphthalene
Carbon tetrachloride	Nitroglycerine
Chlordane	Nonane
Chloroform	Octane
Chrysene	Pentachlorophenol
p-cresol	Phenol
DDT	PCB
Dichlorobenzene	Pyrene
Dichloroethane	Styrene
Dioxane	Tetrachloroethylene
Dioxin	Trichloroethylene
Dodecane	Trinitrotoluene (TNT)
Ethylbenzene	Vinyl chloride
Ethyl glycol	Xylene

Source: Derived from information from Microbac International, *Bioremediation: A Desk Manual for the Environmental Professional*, by Dennis Schneider and Robert Billingsley (Cahners Publishing), and from the *Handbook of Bioremediation*, by Robert S. Kerr (ed.), (Lewis Publishers).

Several firms have developed specific bacteria strains and/or enzymes tailored to the control of pollutants. If the gas stream can be conditioned to provide an environment wherein this bacteria strain or its enzymes can be sustained, the application is a candidate for biofiltration.

Figure 3.2 shows the basic components of an above-ground biofilter. It consists of a preconditioning and humidification chamber to raise the gas relative humidity to 100%, a gas distribution system or header, large vessel containing a mixture of organic material that both supports the bacteria colonies and provides food, the bacteria dosing system, and a condensate return system.

The mixture of organic material the bacteria in the biofilter adhere to is called *biomass*. This biomass may be cellulose or similar wood-based material, peat, carbon (or charcoal), straw, waste organic material, or plastic material (like scrubber packing) or mixtures thereof designed to support the bacteria colonies. Generally, a thin-wetted layer called a *biofilm* is formed throughout this medium, thereby extending the film's surface area (this is akin to the use

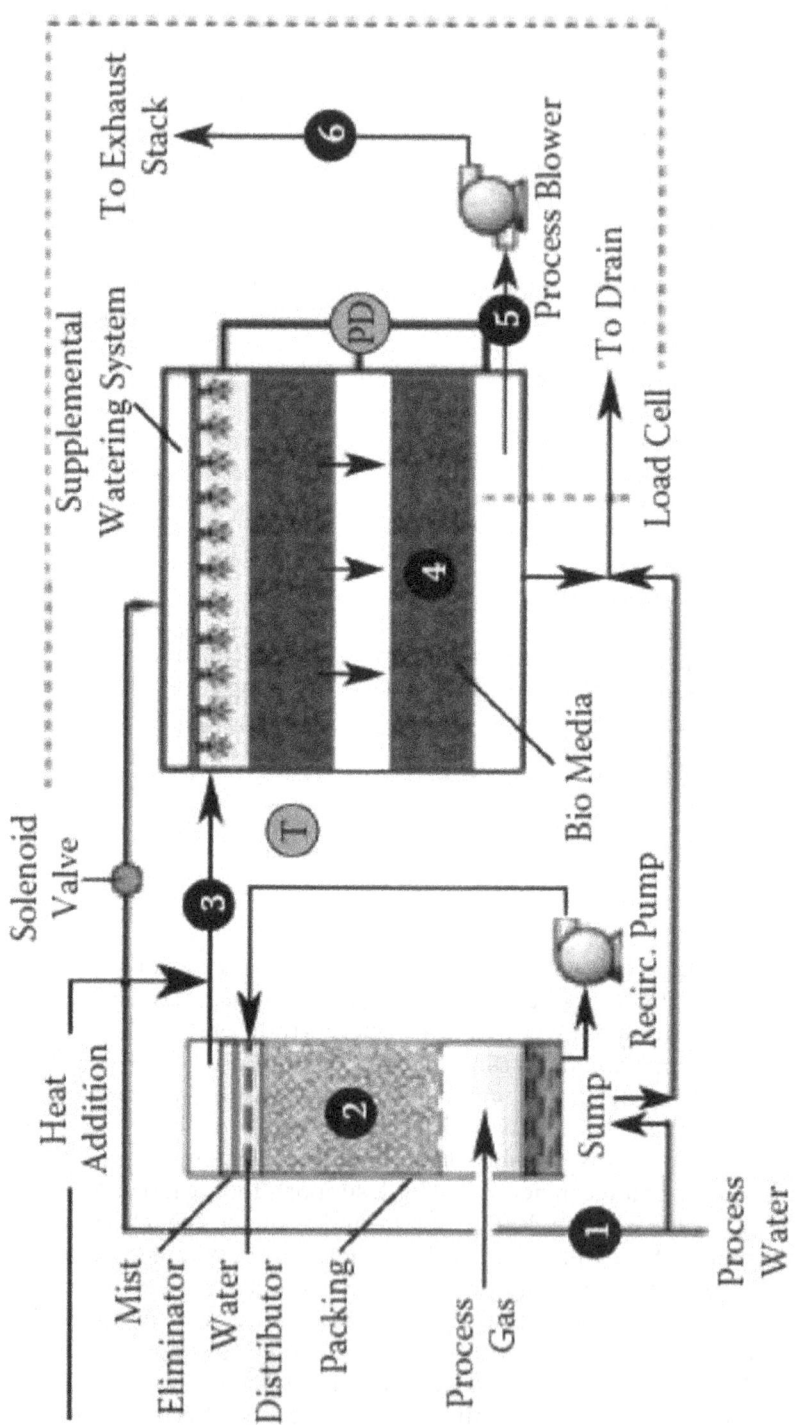

FIGURE 3.2
Biofilter components. (Monsanto Enviro-Chem Systems, Inc.)

of packing in a packed tower). Because the bacteria usually enjoy a warm, moist environment, the humidification spray is used to prevent the biomass from drying out, thereby killing the bacteria. These reactions occur in a moist environment in the biofilm, therefore, the pollutant must be soluble in water and be absorbed.

The contaminant gas is absorbed into the moist biofilm and enzymes secreted by the bacteria reduce or oxidize the contaminant. Given adequate time, the hydrocarbons are converted to carbon dioxide and water vapor. In some cases, they are converted to methane gas, much as in a biologic water treatment system.

In some biofilters, the specific enzyme has been extracted remotely and a concentrated solution of that enzyme is used to coat a supporting medium (such as cellulose gauze). The enzyme fixes oxygen in the air to the hydrocarbon, thereby oxidizing it without depletion of the enzyme itself. In this way, the enzyme is considered to catalyze the oxidation of the contaminant. Figure 3.3 shows a compact gas cleaning device using an enzyme solution supported on a gauze-type medium.

FIGURE 3.3
CAP™ "Clean Air Plant" compact biofilter (SRE, Inc.).

3.4 Primary Mechanisms Used

Absorption is the primary mechanism for the movement of the contaminant from the gas phase into the biofilm. Biologic oxidation occurs using enzymes (called oxygenase). This is the key mechanism for the oxidation of the contaminant once it is absorbed. Enzymes in the bacteria strain act as catalysts to fix oxygen to the contaminant, thereby oxidizing the latter. Some bacteria strains fix other chemicals to the contaminant where a reducing reaction follows. They basically extract a portion of the contaminant, for example sulfur in a mercaptan odor, changing the odorous compound's structure.

For long-chain hydrocarbons, a stepwise cleaving process can occur. Over time, the secreted enzymes break the hydrocarbon chain into smaller components that eventually result in CO_2 and water. These processes occur naturally in the environment. In the biofilter, conditions are created and maintained to make these processes occur more rapidly.

The gases typically mix through diffusion because the gas velocities are exceptionally low (to reduce pressure drop as well). Impaction and interception are minor in a biofilter given the extremely low vapor velocities at which these devices operate.

3.5 Design Basics

In mechanical function, biofilters can be compared to packed towers. The biomass-support medium is the packing, and the biofilm is the absorbing liquid. In the case of biofilters, however, the biofilm is stationary. It is attached to the support medium. The gas, therefore, is caused to move slowly through the biomass so that the contaminant gas can diffuse over to the biofilm surface, and time is allowed for the gas to penetrate the biofilm surface. As a result, gas velocities are under 1–2 ft/s. Biofilters therefore are generally large devices.

They need not be, however, unsightly. Figure 3.4 shows an above-ground biofilter, the housing of which has been designed for function and appearance.

This design is built in modular components to reduce costs and speed installation time. The upper vessel is made from fiberglass-reinforced plastic and is sloped to allow for strength and draining of snow and rain. It sits on a lined concrete basin, which provides structural support and houses the gas distribution system.

Because the bacteria strains used are living organisms, they require a suitable living environment to survive. This usually results in a requirement of humidifying and sometimes heating or cooling the gas stream within a narrow operating window to suit the bacteria strain used. Inlet relative humidity is usually above 95% and the temperatures are 60°F–110°F. Reduced

FIGURE 3.4
Modular biofilter (Envirogen).

moisture can dry out the biomass and excessive temperatures can kill the bacteria. The pH is usually 6–8, although some bacteria strains can function at a pH of 4 to as high as 10. One disadvantage most bioscrubbers have over other technologies described here is the adjustment of pH in the beds. Some bacteria generate acidic byproducts that lower the native bed pH. The acidity could decompose some bed types. This problem gradually increases the differential pressure required to pass gas through the bed and then eventually destroys the bed structure.

The device also must be designed to be replenished. Access doors must be provided, but adequate pull space must also be provided because biofilters are often bulk loaded with biomass support material that is dumped into place and distributed. For this reason, above-ground biofilters often are configured with driveways next to them allowing for mechanical removal and replacement of the substrate into dump trucks or other hauling devices.

3.6 Operating Suggestions

It should be clear from the previous comments that biofilters must be operated within their thermal and humidity window. Care should be taken to provide a reliable supply of humidification water and supply a suitably insulated vessel if cold environments are to be encountered.

For hard water, the use of softened water in the humidification system may be advised to reduce nozzle plugging. If a packed-type humidification device is used, periodic checks should be made regarding the packing condition. The packed zone's pressure drop should be monitored, and the packing replaced if the pressure drop rises above the vendor's prescribed figure.

The condensate from the biofilter should be accumulated and, if recycled, excessively large solids sent through a strainer or filter to prevent nozzle plugging. If the humidification system is lost, the biofilter can be lost. Recycled condensate can also be pH adjusted to partially reduce the effects of acidic buildup in the bed caused by some bacteria.

It is not uncommon with biofilters to experiment with various bacterial cultures and substrates. In part, this may reveal the art side of the science. The reality is that certain bacterial cultures respond to specific pollutants. When a mixture of pollutants is present, problems can result. Patience is therefore an asset if one is trying to tackle a multiple pollutant stream.

It is suggested that the temperature of the post humidification section and the bed temperature should be monitored. The post humidification section should be at the wet bulb temperature or within 2°F–3°F thereof. This indicates near saturation. The bed temperature reflects the bacterial living conditions. The bacteria culture supplier will have a design range within which to operate.

Aside from the service accessibility issues and preconditioning requirements mentioned previously, the biofilter can be operated as any other absorber.

3.6.1 Biotrickling Filter Technology

This device utilizes the same biological mechanism described above in the biofilter section, so it will not be repeated here. This section will point out conditions where this technology has advantages and limitations when compared to biofilter technology.

Devices of this type are freestanding tanks with cone-shaped bottoms. In many applications, the cone-shaped bottoms are buried in the ground. Because these devices have three to five times more packing bed depth than biofilters, they have a longer residence time. This additional residence time increases the interaction between the foul air and the biological media, and that increases the removal efficiency. Furthermore, because the properly maintained media bed used in biotrickling filters has more open space than the media in biofilters, there is less differential pressure required to move the air through the bed (even though the bed is several times deeper) and that saves electrical energy.

Biotrickling filters use recirculated liquid that bathes the entire bed in a uniform way. This provides the ability to regulate the pH of the process in a way that is difficult for biofilters.

3.6.2 Aerobic Digester Technology

There is much literature available on aerobic digester design and operation, so those details are not included here. This section focuses on the innovative applications of aerobic digestion in air quality applications.

Although relatively new as a component in air quality, this technology has been used for over 100 years in wastewater treatment. Pacific Rim Design & Development and other firms successfully integrate this technology with bioscrubber technology and other technologies to treat more effectively a foul air stream that contains mixed organic compounds. The bioscrubber treats the volatile organics with high solubility in waterlike alcohols, and the aerobic digester treats the larger, often less soluble organic compounds like terpenes (organic molecules associated with wood) that require longer residence time for destruction.

The range of compounds that can be treated by aerobic digestion is increased using carefully selected surfactants that increase the solubility of organic compounds without adversely affecting the microbial colony.

The addition of neutral density foam or other material increases the aerobic digester efficiency when it is utilizing biological organisms that prefer to form slimes or colonies. This material is not necessary for organisms with a preferred free-floating habitat.

Particulate material in a waste gas stream will clog most wet scrubbers. This problem is turned into an attribute when an aerobic digester is part of an integrated treatment process and the particulate has a density that is near neutral. As an example, Pacific Rim Design & Development developed a Triple Integrated Process to treat waste gas from engineered wood facilities. In this application, three processes were combined into a single vessel. The waste gas laden with sawdust, alcohols, formaldehyde, and larger organic compounds was first treated with mist that contained surfactants in a downward direction as it passed through a tube at the center of the reaction vessel. This first treatment cooled the gas, agglomerated the particulate material, and enhanced solubility of the larger organic compounds when they reached the aerobic digester. The agglomerated material was forced to drop into an aerobic digester at the bottom of the reaction vessel when the gas made a 180° turn upward into the space between the outer wall of the reaction vessel and the outside of the central tube within the reaction vessel. A bioscrubber in the shape of a ring between the inner tube and the outer wall of the reaction vessel is used to remove the more-water-soluble compounds. Recirculated liquid from the aerobic digester is used to provide moisture and nutrients to organisms in the bioscrubber. The recirculated liquid is filtered prior to being sprinkled over the bioscrubber. The filtered material can be dried and used as fuel in some boilers.

In an integrated system that includes a bioscrubber and aerobic digester, the large thermal mass of an aerobic digester can act as a heat sink to stabilize temperatures in the bioscrubber during shutdowns of short duration.

The heat is distributed to the bioscrubber through recirculated liquid from the bioscrubber.

Integrating more than one biological process into a single treatment has real advantages but also requires a careful study of the habitat requirements for each biological group. This process is most successful when all integrated processes are optimal at the same temperature and pH, and are symbiotic with respect to nutrients, growth patterns, and more.

Aerobic digester design must include a means of adding oxygen to the liquid in its sump. This can be done through the injection of air or oxygen into the sump in a way that promotes mixing. Aerobic digesters used as part of an integrated waste gas treatment process are contained within a vessel, so they must rely on mechanical aeration.

Engineering an integrated system requires a careful study of required residence times and sequences of treatment. This topic is more variable than space is available in this chapter.

3.6.3 Bioscrubber Technology

Just as the biotrickling filter is an enhancement over the biofilter, the bioscrubber is an enhancement over the biotrickling filter. One large difference is that bioscrubbers are contained within a closed vessel and the biofilters and trickling filters are open to the atmosphere. The closed design of the bioscrubber easily allows them to be part of an integrated treatment system. An integrated system can be sequential bioscrubbers, with similar or different microbial environments or an integration with other types of abatement equipment as described under the proceeding section on aerobic digesters. The ability of bioscrubbers to treat in sequential stages also provides them with the ability to treat a wider range of organic materials. This is particularly useful when a waste gas stream has pollutants that are not treated by the same type of organism. In these situations, each stage hosts an organism or group of organisms that effectively target one of several compounds in the waste gas stream.

Bioscrubbers generally have greater removal efficiency than biofilters and biotrickling filters because bioscrubbers have longer residence time for reactions between the waste gas and microbes. In standalone operations (not integrated with other air quality technology), the sump could act as an environment for additional biological treatment. However, the sump is limited by available oxygen in the liquid, so there is limited value in building a sump with more than 50 times the volume of liquid that is being recirculated to the packing bed.

4

Dry Cyclone Collectors*

4.1 Device Type

Dry, cyclonic separators disengage entrained dust from a carrying gas stream. They are often called *cyclone collectors, multicyclones, cyclones, cyclonic separators,* or *cyclonic dust collectors.*

4.2 Typical Applications and Uses

Cyclone collectors are used for product recovery of dry dusts and powders and as primary collectors on high-dust loading (more than 2–5 grs/dscf, or grains per dry standard cubic foot) air pollution control applications.

A common application is the rotary dryer. Used to dehydrate various products from grain to manure, direct- or indirect-fired rotary dryers often use cyclone collectors to capture the entrained dust prior to a secondary collector (such as a Venturi scrubber). The rotating action of the dryer entrains a portion of the product as the product tumbles through the hot, drying air. This product is often valuable in dry form, so the cyclone is used to disengage the dust from the gas stream and be recovered. The residual dust is air conveyed to the downstream device.

Figure 4.1 shows a large-diameter cyclone collector attached to a gas-fired rotary dryer for agricultural product recovery. The cyclone is the large white vessel in the center of the photograph. The large cyclone diameter (in effect increasing the turning radius of the particle path) is used to provide a more-gentle separation of material that would otherwise reduce in size if excessive gas velocities were used. For particulate that resists breakage and size reduction, tighter radius cyclones can be used, resulting in a more compact design. Dryer/cyclone designers select the optimum tangential velocity and turning radius based upon the characteristics of the material to be separated.

Another application is on wood waste or bagasse (sugar cane) boilers where light-entrained ash can be collected in suitably designed cyclones. On wood waste applications, smaller diameter cyclones are often used in "banks"

---

* Portions of this chapter are contributed by Will Duske.

FIGURE 4.1
Rotary dryer and cyclone (Duske Drying Systems).

where each cyclone handles less than 1000 actual cubic feet per minute (acfm) of flue gas. The cyclones are arranged in parallel flow. The smaller diameter cyclones have a shorter turning radius and are more efficient in particulate separation. These are called *multiple cyclone collectors*.

One of the most common uses of cyclones is to protect fans from abrasive dusts. Many dust-producing process applications operate under induced draft. Placing a well-designed cyclone collector ahead of the fan helps protect the latter from abrasive wear and improves the operating life of the fan. If the cyclone alone cannot meet emissions guidelines, another type of device may be used after the fan.

Cyclones are also used to collect trim from paper machines. The edges of the formed paper are trimmed to size using knives and the edge is air conveyed in a continuous ribbon of paper back to a cyclone, then repulped in other equipment, and returned to the paper-making process.

Other uses are sawdust collection, separation of air entrained product from pneumatic conveying systems, primary separation in vacuum cleaning systems, fiber separation, and similar applications where the particulate is heavy enough to be influenced by centrifugal forces.

Dry cyclones are *not* generally used on particulate that is under 5-μm aerodynamic diameter because these size particles (about one tenth the diameter of a human hair) resist inertial separation. Particles under about 5-μm aerodynamic diameter are so small that they lack the required inertial characteristics relative to the carrying gas (usually air) for separation such as a tendency to settle or follow a trajectory and instead are influenced by the movement of the carrying gas itself.

In the past, dry cyclones were often installed with the fan located ahead of the cyclone. In this arrangement, gases and particulate passed through the fan, which was typically used to provide the motive force for ventilating the gases through the system. The fan was therefore located "hot," with the gas stream typically above the saturation temperature, and therefore corrosion issues were often reduced. Since the cyclone was mounted after the fan, the cyclone was considered to be positive, i.e. operating under positive pressure. Of course, the ductwork from the fan to the cyclone and the cyclone itself was under positive pressure, therefore, any leaks were out instead of in. Another downside to moving the particulate through the fan was that the fan wheel and abrasion could in many applications reduce the size of the particulate, thus making the particulate harder to collect.

With the advent of stricter air emissions codes, the trend is to place the fan *after* the cyclone. This places the cyclone under negative pressure, and the possibility that the particulate will reduce in size is less than in the positive mode. Since the cyclone would be running under suction (lower than atmospheric pressure), the cyclone is in a negative mode. Given that the dust in the cyclone is situated at below atmospheric pressure, the dust is typically discharged using rotary locks.

When drying high-density products such as manure, blood meal, bone meal, etc., a primary *drop-out device* such as a shallow-cone cyclonic collector (the shallow cone is used to enhance the removal of the material from the cyclone) can be used. This device reduces the particulate loading to the secondary cyclone. Being a more-lazy design, i.e. lower gas velocities, the larger material tends to separate from the gas stream with less size reduction.

The higher efficiency secondary cyclones are then located after the primary drop-out device on the high-density product applications. On low-density applications, the dryer exhaust could go directly to the higher efficiency cyclone, thereby eliminating the primary separation stage. The higher efficiency cyclones may be configured in single or multiple arrangements depending upon the gas flow parameters, particulate characteristics, and desired efficiency.

Figure 4.2 shows a cyclone installation depicting all of the components mentioned above. The primary cyclone (drop-out device) is in the center of the picture. Note the shallow cone at the base of the device. To the left of the primary collector and elevated is the higher efficiency secondary cyclone. The chamber at the top of the secondary cyclone is where the cleaned

FIGURE 4.2
Cyclone installation on a Duske dryer (Uzelac Industries, Inc.).

gases exit. (Notice the length of the cone at the base of the secondary cyclone.) The gases then move to the induced draft fan, which is located behind the (angled trough) product screw conveyor.

4.3 Operating Principles

One step up of the "complexity ladder" from settling chambers is the family of dust separation devices known as cyclone collectors. These devices primarily use centrifugal force (inertial separation) to spin the entrained particulate from the carrying gas stream. To a lesser extent, they can be considered to be settling chambers wrapped in a cylindrical shape to save space.

The gas stream is typically directed into a cylindrical portion of the device so that a spinning motion is created and sustained for a required number of turns or revolutions to achieve the desired separation. Some designs use a single tangential gas inlet; others use fixed vanes that impart rotational forces to the gas stream. As the gas spins (Figure 4.3), the higher specific gravity dust

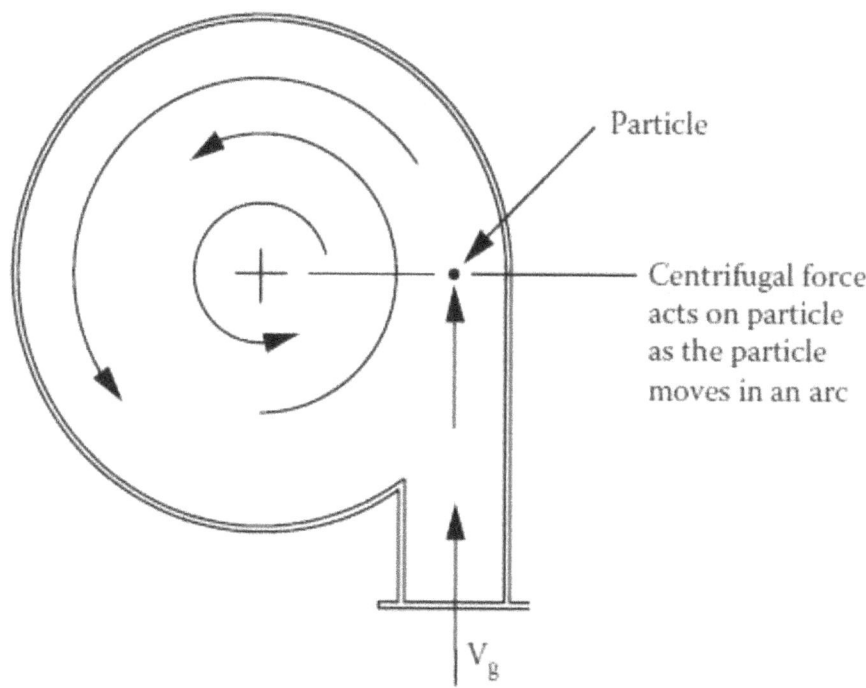

FIGURE 4.3
Cyclonic separation.

is thrown outward toward the containing vessel wall, where it accumulates and slides down the wall surface into a receiving chamber, usually a hopper or other essentially quiescent zone, where the dust accumulates out of the moving gas stream. The dust is usually discharged through a trickle valve or motorized air lock/feeder that prevents air leakage or re-entrainment while allowing the dust to exit.

The typical cyclone includes the following components as seen in Figure 4.4: A tangential gas inlet is used (sometimes incorporating a curved "involute" section) to gradually direct the gas stream for smooth tangential release into the cyclone body. The cyclone body itself is typically a vertical-walled cylinder. The tapered hopper and disengaging section are used to accumulate and separate the dust. The vortex finder (or gas outlet tube) is used to control the ascending vortex. The outlet involute is used to increase the radius of rotation and slowly slow down the spinning gas stream so that the ascending vortex stability is maintained and the rotary airflow is converted to an essentially linear flow with minimal pressure loss.

In general, the more spin cycles or turns imparted to the gas stream, the greater the separation efficiency. Cyclone collector housings are therefore designed to provide a varying number of spins or turns, depending on the

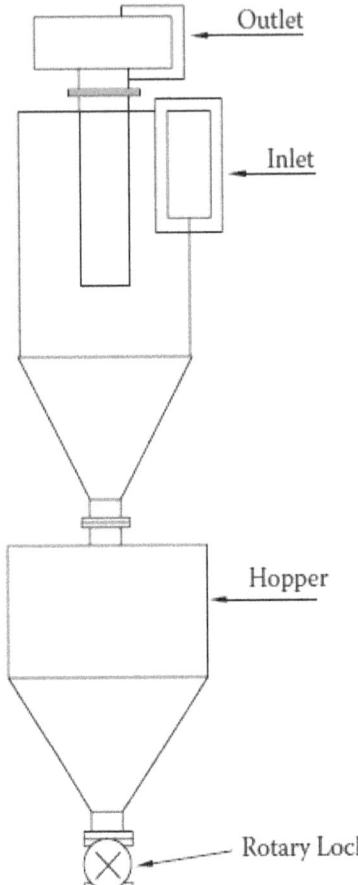

FIGURE 4.4
Basic cyclone collector components.

application. Many dry cyclone collectors use a disengaging hopper to separate the collected material from the gas stream. The hopper is shown at the
lower portion of the drawing in Figure 4.5.

A limiting factor, however, is the friability of the particles (dust) themselves.
A highly friable dust is one that easily breaks down into smaller, more difficult to collect, dust particles as they rub together. Because a cyclone collector
inherently throws the dust close together near the vessel wall, the interaction
between the particles becomes critical in the design. A limit can be reached
wherein the spinning of the dust stream and the friable nature of the dust
achieves equilibrium, and no more dust can be separated.

Because an inertial force is used (centrifugal force and its reaction force), the
particles most influenced by cyclonic action are quite large. Generally, low-
friability particles over 5-μm aerodynamic diameter may be best separated

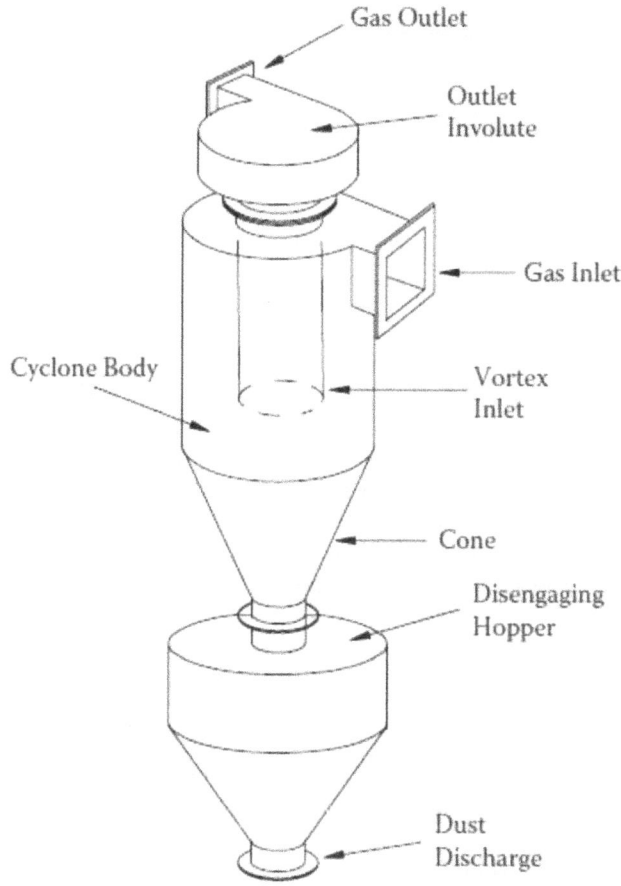

FIGURE 4.5
Dry cyclone (Bionomic Industries, Inc.).

using centrifugal force. Sand particles, for example, are relatively easy to separate with a dry cyclone. Starch, fly ash, and other powders that tend to reduce in size are more difficult to remove with cyclones alone, therefore, cyclones are often followed by additional pollution control devices on those applications.

4.4 Primary Mechanisms Used

Centrifugal force and, to a lesser extent, settling are the forces used in cyclone collectors.

Contradicting forces and effects are same-charge electrostatic forces that could inhibit separation as well as the friability of the particles themselves. Some particulate acquires a charge as it passes through ductwork or a cyclone (piezoelectric effect), thereby making separation more difficult. If the particulate or dust becomes reduced in size, it is more difficult to collect because the effective centrifugal force applied to the particle is a function of its mass.

4.5 Design Basics

Cyclone collectors can be grouped into two general types. The first is the conventional *dry cyclone* and the second is the *multicyclone*. The former is characterized by a relatively large housing with a tangential gas inlet and top central gas outlet, and the latter usually is configured with multiple rows of identical, individual, smaller diameter cyclones. The cyclones of the multicyclone collector are often made from castings, whereas the conventional dry cyclone is usually made from sheet or plate metals.

The conventional dry cyclone is a relatively simple device. Experience has shown that keeping them simple is the best formula for success. To accommodate various gas volumes, they are often grouped in pairs, quads, or even greater numbers.

The gas inlet velocity is usually at or above the conveying velocity of the particular dust being separated. Velocities of 40–65 ft/s are common. The inlet is often rectangular in shape so that the gas enters in wedge form at the tangent line of the cyclone. The width of the inlet is approximately one half the height of the inlet. If the dust is highly friable or abrasive, a velocity toward the lower velocity range is used. If the dust is both heavy and abrasive, a higher velocity must be maintained, so wear plates or even refractory linings are suggested at the gas inlet. The cylindrical body tube length in part dictates the number of turns, and the turning radius (tube diameter) controls the centrifugal force created at a given gas velocity. The higher the gas tangential velocity, the greater the number of turns, the higher the centrifugal force, and the greater the separation.

The cylindrical body length is usually two to three times the body diameter.

The gas outlet velocity is usually 55-65 ft/s and sometimes higher. This vortex finder or outlet tube usually extends down into the cylindrical body portion far enough to prevent dust from short-circuiting from the gas inlet to the outlet tube. An ascending vortex is formed in this tube that turns opposite in direction to the inlet spiral. On cyclones with high-tangential inlet velocities (greater than 100 ft/s), the outlet tube can also be equipped with turning vanes that control the gas swirl. The gas outlet diameter is often approximately one half the cylindrical vessel diameter. Care is taken to avoid having the outlet tube extend down too far into or near the conical section of the collector. If it does, dust near the wall will be drawn back up the outlet tube, lowering the efficiency. The outlet tube length is usually about 1.2–1.5 times the height of the gas inlet.

The tapered or conical portion of the cyclone should be smooth. It is usually made using multiple brake settings if made of metal. If the taper is dented or bumpy, re-entrainment of dust can occur. The taper usually has an angle of at least 60 degrees from the horizontal side. This angle exceeds the angle of repose of most dusts, therefore, bridging at the dust outlet can be reduced.

The gas outlet tube is sized for the expected dust flow rate and allows for a dust velocity of about 4–8 ft/s.

Multicyclone collectors are sized in a similar manner, however, a series of standard tubes are used. Each tube is designed for a given cubic feet per minute of gas flow, and then multiple rows are used to accommodate the design gas flow. Tube volumes of 500–1000 acfm each are common. This results in tubes of 9- to 12-inch inside diameter for many applications. Figure 4.6 shows

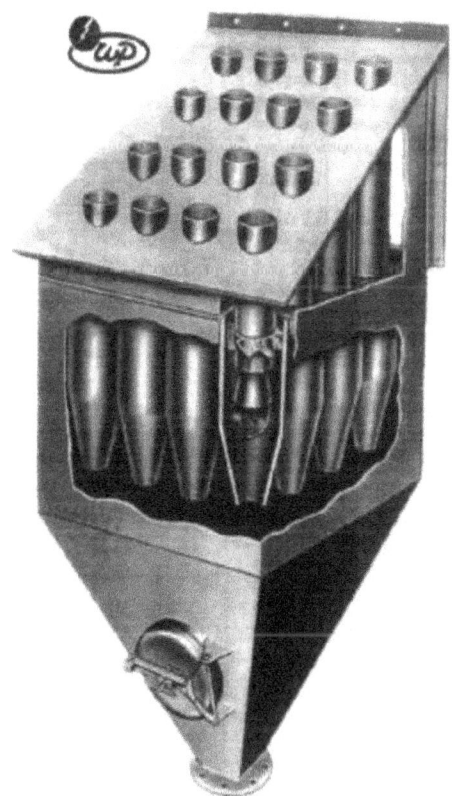

FIGURE 4.6
Multicyclone (Allen-Sherman-Hoff).

a multicyclone collector in cutaway. Notice that the tubes are mounted on a flat tube sheet and the outlet tubes are of varying length. The gas enters from the back of this particular view and exits out the top.

You can also see the vane section. Figure 4.7 shows this more clearly. The vanes look much like a turbine vane and are either cast as part of the tube or are separate pieces fitted into the tube. Quite often, a gas outlet vane is also used to enhance separation and to discharge the finer dust separated in the gas outlet tube.

The multiple cyclone collector works by causing the contaminant particle to move at high speed constrained by the limited radius of the individual tube.

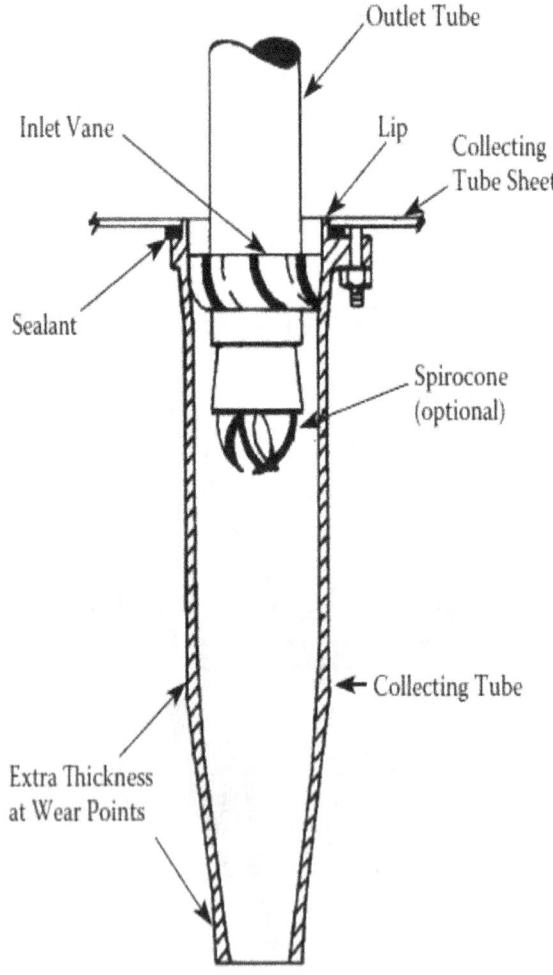

FIGURE 4.7
Components of typical tube (Allen-Sherman-Hoff).

The centrifugal force moves the particle to the tube surface, where it accumulates and drops by gravity down to the collecting hopper. To reduce short-circuiting of dust in the tube, a special outlet tube is used, often with vanes that impart a rotation to the ascending gas stream. Figure 4.8 shows the basic operating principles of the multiple cyclone.

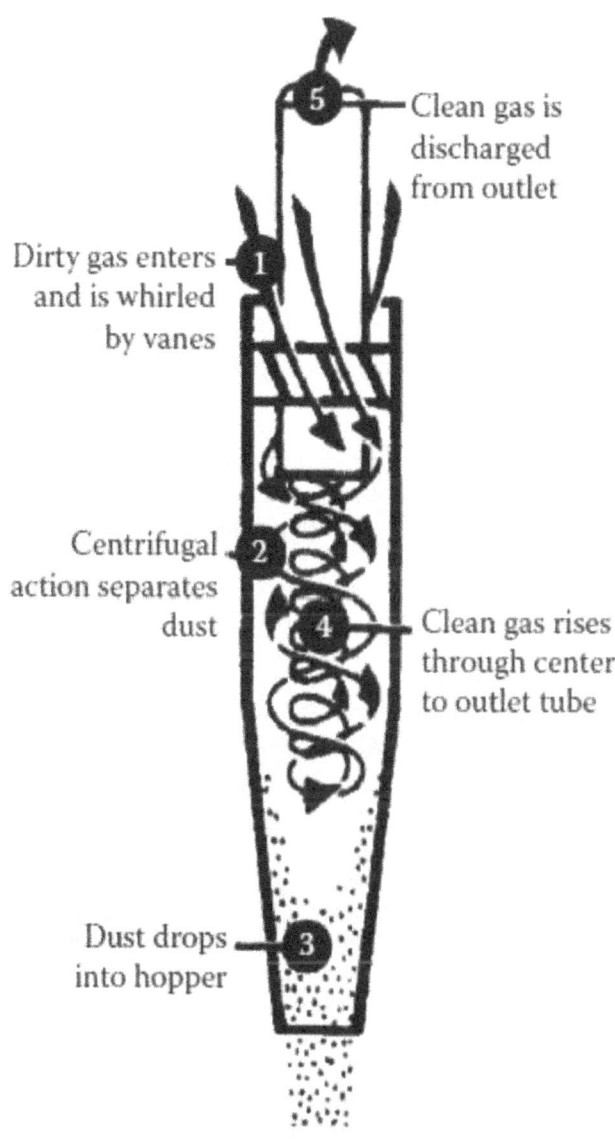

Clean gas is discharged from outlet

Dirty gas enters and is whirled by vanes

Centrifugal action separates dust

Clean gas rises through center to outlet tube

Dust drops into hopper

FIGURE 4.8
How a multiple cyclone works (Allen-Sherman-Hoff).

4.6 Operating/Application Suggestions

The proper application of a cyclone collector starts with a knowledge of the type of dust being collected and its concentration.

Cyclone designers have accumulated data on a variety of dusts and their characteristics. Some dusts are spherical and others are fissured. Some particulate is oblong in shape. These characteristics affect their collectability using centrifugal force.

To make life easier, particulate is often characterized by its aerodynamic rather than physical diameter. The aerodynamic diameter can be considered to be its real-world effective diameter versus its actual physical characteristics as they would appear in, for example, a photograph. The aerodynamic diameter is obtained using a particle sizing device such as the cascade impactor, which separates particulate by size in accordance with its aerodynamic behavior.

Cyclone collector designers use the aerodynamic characteristics and loading to select the appropriate cyclone(s). If the dust loading is very high and the particulate is friable, for example, the designer may use a larger diameter cyclone with reduced turning radius. Often, cyclones are used in stages or groups where the gas flow is split into multiple streams and the separation conducted under more controlled conditions where the dust layer at the wall is thinner. Figure 4.9 shows a sketch of a multiple or dual cyclone. These units often share a single dust collection hopper and single rotary lock or discharge valve.

On multicyclone units, a condition called *hopper recirculation* can occur that reduces efficiency. When this condition exists, some dust-laden air goes into the inlet tube of one cyclone and short-circuits up the discharge tube of another cyclone. The telltale sign is usually an accumulation of dust immediately above the offending tube's gas discharge pipe. This often occurs when the defective cyclone's inlet vanes are broken or if the tube housing itself fails. When inspecting the interior of a multicyclone collector, look for these deposits. The offending tube can often be replaced without affecting its neighbors.

To allow the dust to exit the collecting hopper, a valve must be used to allow dust out but keep air from entering. Trickle valves and rotary locks are commonly used for this service (some cyclones in batch operation service use drum fittings that seal directly onto a receiving drum).

Trickle valves have counterbalanced plates inside of their housings that allow a measured weight of dust to discharge without allowing gas to enter or escape. One such trickle valve is shown in Figure 4.10. Counterweighted double-stage valves are also often used to reduce air infiltration. The external counterweight applies pressure to an internal plate that seals in the dust until the weight of the dust above the plate is sufficient to overcome the force of the

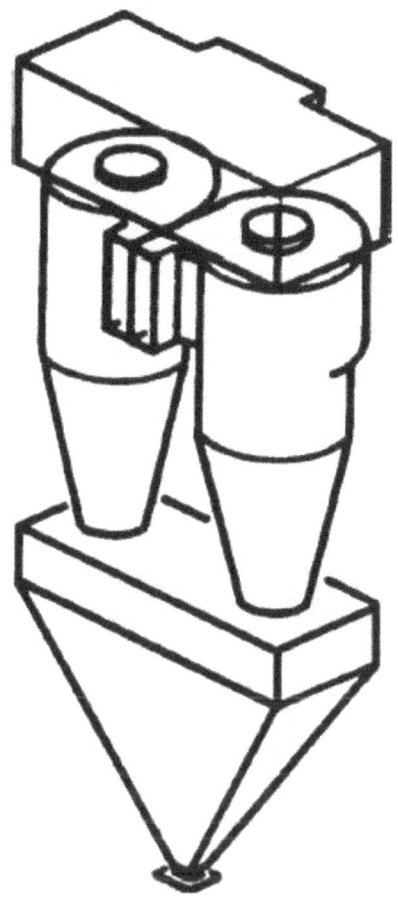

FIGURE 4.9
Dual cyclone on common hopper (Bionomic Industries, Inc.).

counterweight. The dust momentarily caught between the counterweighted plates acts as an additional sealing medium. These trickle valves may be *single dump* (one sealing chamber) or *double dump* (two sealing chambers) as shown in Figure 4.10. Sometimes, the discharge dust exits to a collecting drum or conveyor. Fabric filter or wet-type dust connection equipment is often used to collect and control fugitive dust in the particulate discharge area.

The rotary lock is another commonly used device to allow the uniform discharge of captured dusts from dry cyclones, dry precipitators, and other dry collection devices. Figure 4.11 shows such a rotary lock. These designs have evolved over the years to provide very effective discharge of collected dusts. The figure shows the lock with its drive mechanism (usually a gear reduction motor) removed.

FIGURE 4.10
Trickle valve (photo courtesy of the Wm. W. Meyer Co.).

FIGURE 4.11
Rotary lock (photo courtesy of Wm. W. Meyer Co.).

The inlet flange is shown at the top and the outlet flange is at the bottom. The dust flows vertically downward. The vanes are driven slowly by the rotation of the shaft shown to the right of the photo. Typically, a high-torque rotary motor or actuator is applied to the keyed shaft. At any given time, there is no direct passage of dust out or air in. These rotary locks operate much like the revolving doors used in hotels and other buildings that allow people and things to enter and leave with minimal air entry or loss. As with the trickle valve, the rotary lock may discharge into a collecting drum or conveyor.

A very common problem for any cyclone is excessive gaps in the rotary lock or trickle valve, allowing ambient air to be drawn into the cyclone hopper (if the cyclone operates under induced draft), which acts as an air lift to jettison the dust out of the collector. To mitigate this, rotary locks using adjustable end plates and rotor seals are used. With use of a motorized air lock, the end plate is constantly pressed against the rotating sealing vane in the device, thus reducing air leakage. These locks should be periodically inspected and adjusted as required but are often neglected given the dusty environment in which they must operate. It is literally a dirty job but someone has to do it.

Most cyclones must be installed vertically because their operation relies on stable, controlled vortices that Mother Nature (and the laws of physics) tells us operate best vertically. The ascending vortex that forms should be symmetric with the cyclone body otherwise, an imbalance in the rotational forces can occur. If this happens, dust from the cyclone body can be vacuumed out the outlet tube, causing a reduction in efficiency.

Another common problem is using an elbow immediately after the cyclone. The elbow can sometimes upset the spinning action of the ascending vortex, causing an imbalance. At least two diameters of straight ductwork at the cyclone discharge before the elbow usually solve the problem. Dry cyclones often use the involute-type outlet box to reduce these imbalances and produce a stable ascending vortex.

Excessive dust levels in the cyclone hopper can cause serious problems. Consider the spinning gases as a confined tornado inside the vessel. If you let the tornado (spinning vortex) touch down on the accumulated dust, the vacuuming action can lift the dust up and out of the collector. The vortex flow can often be reduced by the addition of a vortex breaker located above the trickle valve or air lock. Some facilities use bin detector devices to monitor the hopper dust level and actuate the rotary lock or trickle valve (above) to keep the level low enough to prevent touchdown.

For highly abrasive dusts, replaceable inlet scroll-wear plates are often used. Made of abrasion-resistant plate, they help reduce the erosive effects of such particulate.

In multicyclone collectors, a dust recirculation pattern can occur inside the cyclone modules. Gas (and dust) can migrate from tube to tube depending upon the differential pressure across the tubes. This can be mitigated by increasing the differential pressure between the hopper chamber and the clean air plenum. This is accomplished by pulling a draft on the hopper air space and directing the gas flow to a baghouse or other external dust-collecting devices.

Properly designed cyclone collectors are effective devices for the recovery of dry products and as primary collectors for subsequent additional air pollution control stages.

5

Electrostatic Precipitators

Bob Taylor

BHA Group, Inc., Kansas City, Missouri

5.1 Device Type

Electrostatic precipitators are used for the purpose of removing dry particulate matter from gas streams. They basically apply an electrostatic charge to the particulate and provide sufficient surface area for that particulate to migrate to the collecting plate and be captured. The collecting plates are rapped periodically to disengage the collected particulate into a receiving hopper.

5.2 Typical Applications and Uses

Dry electrostatic precipitators are used to remove particulate matter from flue gas streams exiting cement kilns, utility and industrial power boilers, catalytic crackers, paper mills, metals processing, glass furnaces, and a wide variety of industrial applications (see Figure 5.1).

An electrostatic precipitator is a constant pressure drop, variable emission particulate removal device offering exceptionally high particulate removal efficiency.

There is a unique jargon involving electrostatic precipitators. If you contemplate purchasing or studying the use of one, perhaps the following buzzword list will prove helpful. It is in alphabetical order, so if you see a word that you do not understand, just jump down the list to find the offending word.

> *Air splitter switch:* An air splitter switch is mounted at the high-voltage bushing contained on the transformer rectifier. The purpose of the switch is to isolate one of the two electrical sections served by the transformer rectifier while the other operates.
> *Antisneak baffle:* A deflector or baffle that prevents gas from bypassing the treatment zone of the precipitator.

FIGURE 5.1
Typical electrostatic precipitator in operation (BHA Group, Inc.).

Arc: Arcs occur within the high-voltage system because of uncontrolled sparking. If measurable current flow is detected, damage will occur to internal components.

Aspect ratio: The treatment length is divided by treatment height. A higher number is more favorable for collection efficiency.

Back corona: Occurs in high-resistivity dust applications. As a result of the dust resistivity, a voltage drop occurs across the layer of dust on the collecting plates. The application of current to the field builds the charge on the surface of the dust layer until the breakdown voltage of the dust is achieved. At this point, a surge of current occurs from the surface of the dust to the collecting plate, causing localized heating of the dust. The dust explodes back into the gas stream carrying a charge opposite of the electrons and gaseous ions. This causes collection efficiency to degrade and dust reentrainment to increase.

Bus section: Smallest isolatable electrical section in the precipitator.

Casing: Gas-tight enclosure within which the precipitator collecting plates and discharge electrodes are housed.

Chamber: Common mechanical field divided in the direction of gas flow by a partition. The partition is either a gas-tight wall or open structural section.

Cold roof: This is the walking surface immediately above the hot roof section.

Collecting surface: Component on which particulate is collected. Also known as collecting plate or panel.

Corona discharge: The flow of electrons and gaseous ions from the discharge electrode toward the collecting plates. Corona discharge occurs after the discharge electrode has achieved high enough secondary voltages.

Current-limiting reactor: This device provides a fixed amount of inductance into the transformer rectifier circuit. Some current-limiting reactors have taps that allow the amount of inductance to be varied manually when the circuit is not energized.

Direct rapping: Rapping force applied directly to the top support tadpole or lower shock bar of a collecting plate.

Discharge electrode: The component that develops high-voltage corona for the purpose of charging dust particles.

Disconnect switch: A switch mounted in the high-voltage guard or transformer rectifier that allows the electrical field to be disconnected from the transformer rectifier.

EGR: Electromagnetic impact gravity return rapper used for cleaning discharge electrodes, collecting plates, and gas distribution devices. An electromagnetic coil when energized raises a steel plunger, which can free fall onto the rapper shaft after the coil is deenergized.

Electrical bus: The electrical bus transmits power from the transformer rectifier to each electrical field. Generally fabricated from piping or tubing.

Electrical field: An electrical field comprises one or more electrical sections energized by single transformer rectifier. A single voltage control serves the electrical field.

Gas distribution device: A gas distribution device is any component installed in the gas flow for the purpose of modifying flow characteristics.

Gas passage: The space defined between adjacent collecting plates.

Gas passage width: The distance between adjacent collecting plates. Consistent within a mechanical field but can vary between fields contained in a common casing.

Gas velocity: Gas velocity within a precipitator is determined by dividing total gas volume by the cross-sectional area of the precipitator.

Ground switch: A device mounted in the high-voltage guard or the transformer rectifier for the purpose of grounding the high-voltage bus. This does not disconnect the field from the transformer rectifier.

High-voltage guard: High-voltage guard surrounds the electrical bus. Generally fabricated from round sections that provide adequate electrical clearances for the applied voltages.

High-voltage support insulator: The ceramic device fabricated from porcelain, alumina, or quartz that isolates the high-voltage system from the casing. Typically, a cylindrical or conical configuration, but some manufacturers use a post-type insulator.

Hopper: A casing component where material cleaned from the discharge electrodes and collecting plates is collected for removal from the system. Can be pyramidal, trough, or flat bottom.

Hot roof: Comprises the top gas-tight portion of the casing.

Insulator compartment: An enclosure for a specific quantity of high-voltage support insulators. Typically contains one insulator but may contain several. The insulator compartment does not cover the entire roof section.

Key interlock: A key interlock system provides an orderly shutdown and startup of a precipitator electrical system. A series of key exchanges connected to de-energizing equipment eventually provides access to the internals of the precipitator.

Lower frame stabilizer: A lower frame stabilizer controls electrical clearances of the stabilizer frame relative to the mechanical field. This device typically contains an insulator referenced to the hopper, casing, or collecting plate and attached on the other end to the stabilizer frame.

Mechanical field: This is the smallest mechanical section that comprises the entire treatment length of a collecting plate assembly and extends the width of one chamber.

Migration velocity: The velocity at which the particulate moves toward the collecting plate. Measured in either feet per second or centimeters per second.

Normal volume: This is the normalized condition when using metric measurements.

Opacity: An indication of the amount of light that can be transmitted through the gas stream. Measured as a percentage of total obscuration.

Partition wall: Divides adjacent chambers in a multiple chamber precipitator. Can be gas tight, but also can be a row of supporting columns.

Penthouse: An enclosure that houses the high-voltage support insulators. Typically covers the entire roof section of the precipitator casing. This is a gas-tight enclosure that cannot be entered when the precipitator is operating.

Perforated plate: A perforated steel plate, typically 10 gauge, that is placed perpendicular to gas flow for the purpose of redistributing the velocity pattern measured within the precipitator. The perforation pattern is typically not uniform across the panels providing specific flow patterns.

Primary current: The current provided at the input of a transformer rectifier. It will be measured in alternating current (AC) amps.

Primary voltage: The voltage provided at the input of a transformer rectifier. It will be measured in AC volts.

Purge heater system: Intended to provide heated, pressurized, and filtered air into the insulator compartments or penthouse. An electric heater element or sometimes steam coil heats air that has been drawn through a filter by a blower. The conditioned air is then distributed into the support insulators.

Rapper: A device responsible for imparting force into a collecting plate or discharge electrode for the purpose of dislodging dust.

Rapper insulator shaft: An insulator shaft that isolates the high-voltage rapping system from the casing. Can be fabricated from any material with high dielectric, but typically porcelain, alumina, or fiberglass-reinforced plastic.

Rigid discharge electrode: A discharge electrode that is self-stabilizing from the high-voltage frame down to the stabilizer frame. Typically constructed from tubular or roll-formed material. Individual emitter pins or other corona generators are affixed to the surface for the purpose of generating high-voltage corona.

Rigid frame: Rigid frames are associated with tumbling hammer-type precipitators. A rigid frame that encompasses the entire gas passage area is provided for the purpose of support of individual discharge electrodes.

Saturable core reactor: Sometimes also called an SCR, this is an antiquated method of providing inductance into the transformer rectifier circuit. The saturable core does vary impedance but is extremely slow to react and introduces distortion into the waveform. Replaced by the current-limiting reactor.

Secondary current: Current measured at the output side of a transformer rectifier. It will be measured in direct current (DC) milliamps.

Secondary voltage: Voltage measured at the transformer rectifier output. It is measured in DC kilovolts.

Silicon control rectifiers: Silicon control rectifiers are the switches that control power input to the electrical field. The voltage control turns the silicon control rectifier on and off based on the sparking occurring within the field.

Spark: A spark within a precipitator occurs between the high-voltage system and the grounded surfaces. There is a minimum of current flow during a spark, and as a result internal components are not damaged. Sparking is the method by which voltage controls determine the maximum usable secondary voltage that can be applied to an electrical field.

Specific collecting area: Specific collecting area (SCA) is the total amount of collecting plate area contained in a precipitator divided by the gas volume treated. When referenced to a common gas passage width, values for SCA can be compared to define relative capability of precipitators.

Transformer rectifier: A device to rectify the AC input to DC and step up the voltage to the required level. A single voltage control serves each transformer rectifier.

Treatment length: Total length of all mechanical fields in the direction of gas flow.

Treatment time: Treatment time or retention time is calculated by dividing the treatment by the gas velocity.

Tumbling hammer rapping: A rapping system utilizing a series of hammers mounted on a shaft common to a mechanical field. When the shaft rotates or drops, the hammers strike an anvil connected to the collecting plates or high-voltage frames.

Turning vane: Turning vanes are installed within ductwork or precipitator inlet and outlet transitions to direct flow to a specified position.

Voltage control: A voltage control serves a single transformer rectifier for the purpose of maximizing power input to the electrical field that it serves.

Weather enclosure: This is a weatherproof enclosure over the top of a precipitator for the purpose of facilitating maintenance during adverse weather. It is not for the purpose of isolating high-voltage electrical sections.

Weighted wire: A discharge electrode fabricated from wire that is tensioned by a cast iron weight.

To make sense of these terms, the following illustrations indicate some of the terms for standard configuration electrostatic precipitator components: Figure 5.2 shows a complete electrostatic precipitator. The cutouts show specifics that will become clearer. The details shown will become more obvious

FIGURE 5.2
Complete electrostatic precipitator (BHA Group, Inc.).

FIGURE 5.3
Exploded detail of single field (BHA Group, Inc.).

as we look more deeply at selected components. Figure 5.3 shows better detail of a single field. Note the detail of the rapper tranes. The rappers that clean the collecting plates are configured differently from those for the high-voltage system. The collecting rapping system is shown in Figure 5.4, and the high-voltage rapping system is shown in Figure 5.5.

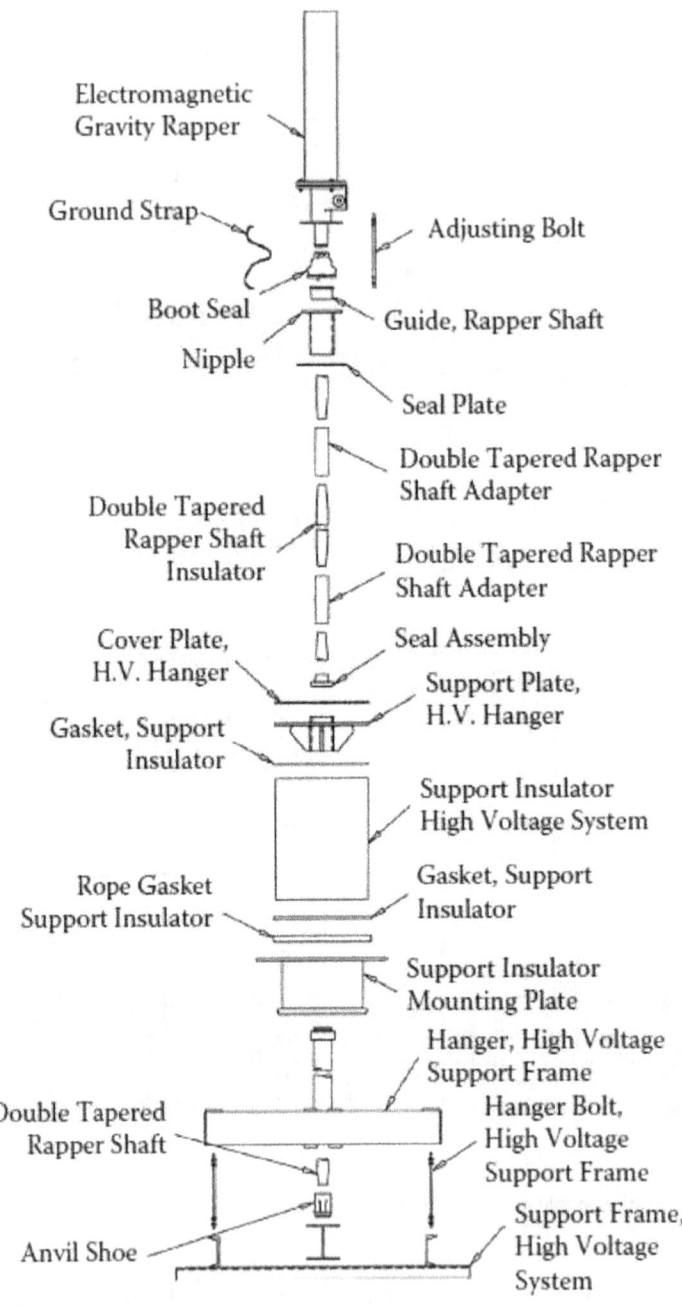

FIGURE 5.4
Collecting system components (BHA Group, Inc.).

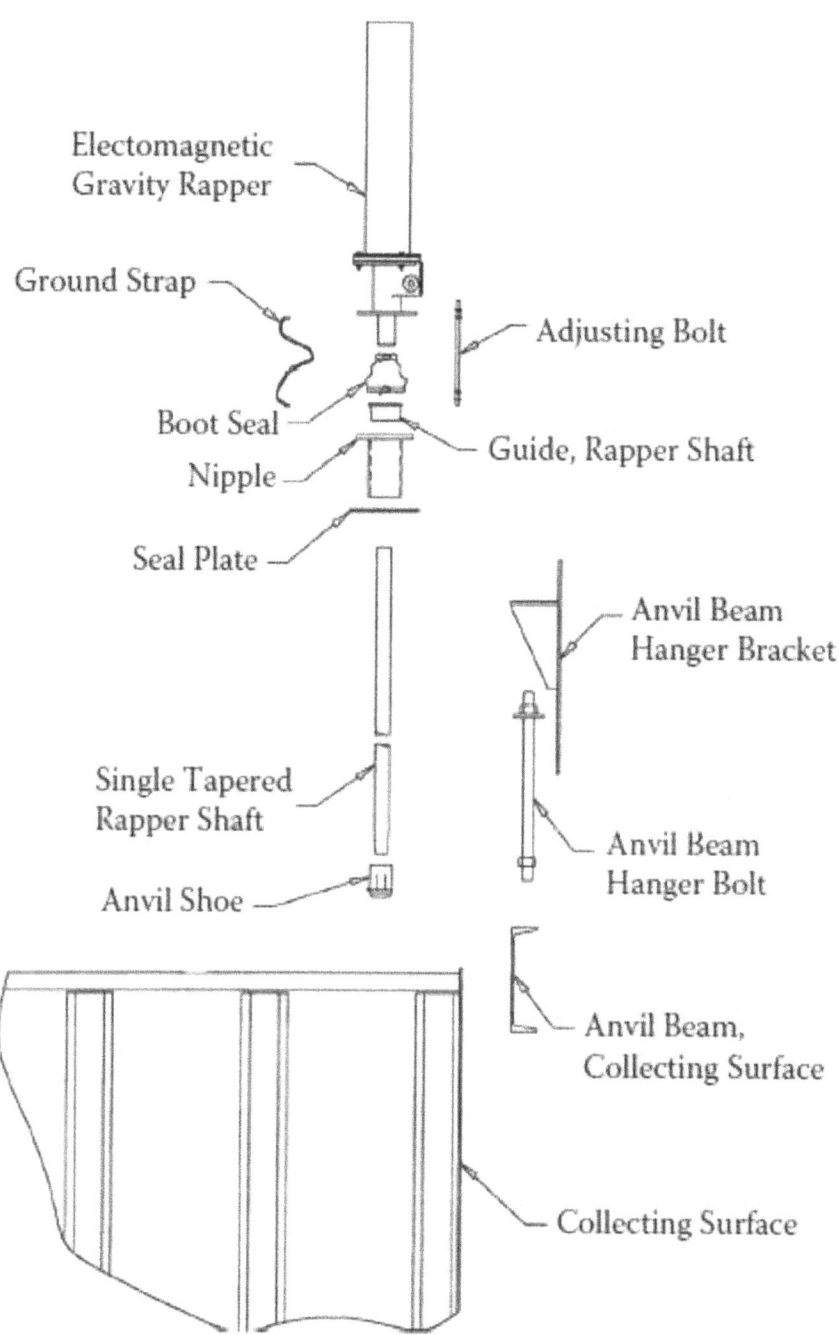

FIGURE 5.5
High-voltage system components (BHA Group, Inc.).

5.3 Operating Principles

The basic principle of an electrostatic precipitator is to attract charged dust particles to the collecting plates where they can be removed from the gas stream.

Dust entering the precipitator is charged by a corona discharge leaving the electrodes. Corona is a plasma containing electrons and negatively charged ions. Most industrial electrostatic precipitators use negative discharge corona for charging dust.

When charged, the dust particles are driven toward the collecting plates by the electromagnetic force created by the voltage potential applied to the discharge electrodes. An electrostatic precipitator contains multiple mechanical fields located in series and parallel to the direction of the gas flow. Each mechanical field comprises a group of collecting plates that define a series of parallel gas passages. These passages run in the direction of gas flow. Bisecting the gas passage is a series of discharge electrodes, also running in the direction of gas flow.

A mechanical field contains one or more electrical fields. A single transformer rectifier serves each electrical field. Multiple electrical sections can be contained in a single electrical field.

Some form of mechanical cleaning device serves both the high-voltage and collecting system. These rappers can take the form of hammers mounted on a drive shaft, externally mounted pneumatic rappers, or electromagnetic impact devices. The basic intent is to impart a mechanical force to the collecting plates and discharge electrodes to cause dust to drop to the bottom of the precipitator for disposal.

During operation, AC is applied to the voltage control cabinet. Inside the cabinet is a voltage control and silicon control rectifier. The voltage controls the flow of current through the silicon control rectifier. Current from the silicon-control rectifier enters the current-limiting reactor, then the transformer rectifier. The current-limiting reactor serves to reduce distortion in the AC waveform and limit current flow during sparking. The transformer rectifier takes the AC and converts it to DC. In addition, the primary voltage is stepped up to significantly higher secondary voltages. Typical secondary voltages are in the range of 45,000–115,000 kV. Current exiting the transformer rectifier enters the electrical field where charging occurs.

Based on data measured within the electrical field, the voltage controls fire the silicon-control rectifier to introduce current into the field. The amount of time that current is applied to the field is a function of the voltage at which sparking occurs within the field. When a spark is detected within the electrical field, the voltage quenches the spark by turning power off or reducing power levels to a preset level. Once the quenching period is satisfied, the voltage control ramps up power applied to the field in search of the next spark.

5.4 Primary Mechanism Used

As indicated, dust must be charged to be attracted to the collecting plates. This charging occurs between the collecting plates where the discharge electrodes are located. The presence of charge in the gas passage is a function of the secondary voltage applied to the electrical field.

5.5 Creation of Charge

Applying secondary voltage to the discharge electrodes creates the corona discharge. The minimum secondary voltage at which current flow is created is called the corona onset voltage. Typical corona onset voltages range from 12,000 to 25,000 V. In general, the corona onset voltage is a function of the discharge electrode geometry, process gas characteristics, and dust characteristics. If the electrical field operates at a secondary voltage lower than the corona onset voltage, no charging will occur.

Two basic charging mechanisms occur within an electrostatic precipitator: field and diffusion charging. Particle size has a major impact on the type of charging that occurs. A discussion of each mechanism follows.

5.5.1 Field Charging

This charging mechanism generally dominates in particles of 1.5 μm and larger. Dust particles intercept negative ions and electrons emanating from the discharge electrode. Charge physically collects on the surface of the dust, reaching a saturation point. This type of charging is very rapid, occurring in the first few feet of the precipitator.

5.5.2 Diffusion Charging

Particles less than 0.5 μm in diameter are charged using a diffusion mechanism. Diffusion charging is the result of commingling of particles and charge contained in the gas stream. Charging follows the pattern of Brownian movement is a gas stream; charge does not accumulate on the dust but acts upon it. This mechanism of charging is slow compared with field charging.

As seen from the explanation, neither of the two charging mechanisms dominates when particle diameter is between 0.5 and 1.5 μm. In this size range, the combination of field and diffusion charging occurs with neither mechanism dominating. As a result, the combined charging occurs at a rate much slower than either of the two mechanisms. When a precipitator experiences a dominant quantity of particles in this size range, performance is suppressed.

5.6 Design Basics

The relationship between operating parameters and collection efficiency is defined by the Deutsch-Anderson equation. There are several modifications to the original formula, but the basic equation is:

$$\text{Efficiency} = e^{-(A/V)*W}$$

where:

W $\quad = (E_0\, E_p\, a/2\, \pi\, \eta)$
Efficiency = Fractional percentage collected from gas stream
A $\quad$ = Total collecting plate area
V $\quad$ = Volumetric flow rate in actual terms
W $\quad$ = Migration velocity of dust toward collecting plates
E_0 $\quad$ = Charging field strength
E_p $\quad$ = Collecting field strength
A $\quad$ = Particle radius
H $\quad$ = Gas viscosity
Π $\quad$ = Pi

The simple explanation of the Deutsch-Anderson equation is that the precipitator collection efficiency is defined by the speed of the dust toward the collecting plates and the amount of collecting plate area relative to the total gas volume.

Increasing the migration velocity of the dust will increase collection efficiency of the electrostatic precipitator. Increasing the amount of collecting plate area available to treat the gas volume will also increase collection efficiency.

Likewise, reductions in migration velocity or plate area, or an increase in gas volume, will cause collection efficiency to decrease.

As shown previously, removal efficiency of an electrostatic precipitator is largely determined by the ratio of the total collecting plate area to the gas volume treated. This ratio is called the SCA. The higher the value for SCA, the greater the removal efficiency for the electrostatic precipitator.

Also critical to precipitator performance is treatment time. Higher treatment time implies a larger precipitator available for gas treatment. This parameter is a function of the total length of the mechanical fields in the direction of gas flow and the velocity of the gas through the precipitator. High-efficiency electrostatic precipitators generally provide treatment times greater than 10 s.

Aspect ratio (treatment length divided by collecting plate height) should be greater than 0.8. If the collecting plate becomes too tall relative to the available treatment length, problems associated with dust distribution and reentrainment will increase.

5.7 Resistivity of Dust

There are two types of conduction characterized in dust: surface conduction and volume conduction.

Dust resistivity plays a major role in defining electrostatic precipitator collection efficiency. It is generally accepted that electrostatic precipitators operate most effectively when dust resistivity is in the range of 5×10^9 to 5×10^{10} ohm-cm.

When dust resistivity drops below this range, the dust releases its charge readily to the collecting surface. As a result, the dust migrates to the collecting plates, where it immediately loses its charge. The charge in conjunction with the cohesive nature of the dust keeps the dust on the collecting plates. If the charge is lost, the dust is likely to be reentrained back into the gas stream. Conversely, high-resistivity dust retains charge for extended periods. When the high-resistivity dust deposits on the collecting plates, charge does not dissipate. In fact, charge continues to accumulate due to the constant corona emanating from the discharge electrodes. As a result, high-resistivity dust is difficult to remove from the collecting plates. It is not uncommon for high-resistivity dust applications to require periodic manual cleaning to restore precipitator performance.

Figure 5.6 indicates relative dust resistivity for varying sulfur content of coal. Similar relationships exist between resistivity and process gas moisture content.

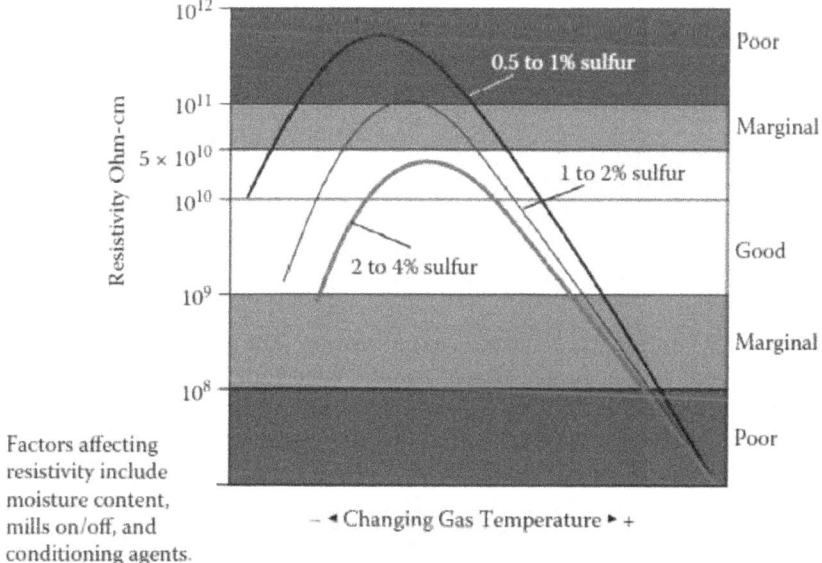

FIGURE 5.6
Average ash resistivity versus gas temperature (BHA Group, Inc.).

Flow of current through the dust layer occurs in one of two methods: surface conduction or volume conduction. The temperature at which the process operates defines the dominant method of conduction.

Volume conduction is the process of current flow *through* the particle. This conduction method occurs on the hot side of the resistivity curve. The hot side starts at the point on the resistivity curve where increasing temperature produces reduced resistivity.

Volume conduction is determined by the resistivity of the constituents at the process operating temperature. Changing the moisture content or adding conditioning agents to the process gas stream will have minimal impact on the hot side of dust resistivity.

Surface conduction occurs on the cold side of the resistivity curve. The cold side is defined as the peak on the resistivity curve toward the slope of decreasing resistivity with decreasing process temperature.

Surface conduction occurs across the surface of the dust particle. Current flow is largely determined by the quantity and type of gasses condensed on the surface of the particle. When operating on the cold side of the resistivity curve, addition of conditioning agents or moisture will generally improve operation.

5.8 Operating Suggestions

Several activities are necessary to ensure effective operation of an electrostatic precipitator.

5.8.1 Air Load/Gas Load Testing

Air load/gas load testing is the process of operating the electrical fields under known conditions. The air load test occurs before startup or immediately after shutdown of the process. Before testing, each electrical field is isolated and confirmed to be ready for energization of the transformer rectifiers. Fans are set at a low flow rate, adequate to provide some ventilation of the electrostatic precipitator.

The voltage control is set in a manual condition. The secondary voltage levels applied to a single electrical field are increased incrementally from zero. At each increment, the measured secondary current is recorded. The secondary voltage at which secondary current is first observed is called the *corona onset voltage*. The secondary voltage is increased to the point at which the nameplate rating of the transformer rectifier is achieved or the field sparks. This process is repeated for each electrical field until all are complete.

As a practical matter, all air load tests should be performed from the outlet electrical field working toward the first field of the precipitator. Sparking generates ozone, which lowers the sparking threshold of a field.

The data derived from the air load test can be plotted to create a volt versus amps (V–I) chart. The air load V–I chart can then be compared with that achieved during operation. Most modern voltage controls contain an automatic air load function that will ramp the voltage and create the plot.

Tests like the air load can be accomplished during operation of the process. These tests are called *gas load tests*. The curve plotted from these process conditions can be used to diagnose electrostatic precipitator operating problems.

5.8.2 Alignment

As indicated, the speed of the dust toward the collecting plates is a function of the applied field strength. The secondary voltage levels achieved largely determine field strength.

It is desirable to have the discharge electrodes centered within the gas passage and between collecting plate stiffeners. As the electrical clearance decreases due to changes in alignment, the voltage at which sparking will occur decreases. Bowed collecting plates, misaligned fields, and foreign objects in the gas passage will increase spark rates and decrease secondary voltage levels.

5.8.3 Thermal Expansion

When the casing and internal components of a precipitator achieve operating temperature, thermal expansion may change the electrical alignment. In this condition, electrical conditions may be acceptable at ambient temperatures but not at operating temperatures.

It is essential to ensure that the components can accommodate growth associated with thermal expansion and still maintain acceptable electrical clearances.

5.8.4 Air In-Leakage

As shown in the Deutsch-Anderson equation, collection efficiency is a function of SCA. If ambient air is leaking into a negative pressure gas stream, the precipitator is forced to treat a larger total gas volume. There are other reasons that air in-leakage reduces precipitator performance.

Ambient air generally contains a lower water content compared with flue gas. As shown in the resistivity section, increasing moisture content improves dust resistivity. When ambient air leaks into the gas stream, the average moisture content is reduced, and resistivity generally increases. This applies to those units operating on the surface conduction side of the dust resistivity curve.

5.8.5 Rapping

The ongoing satisfactory performance of an electrostatic precipitator is a function of maintaining the collecting surfaces and discharge electrodes free from excessive dust layer.

Creation of an acceptable rapping program is an iterative process. There is no formula that establishes the correct program. As changes are implemented to the rapper program, they must be evaluated in terms of their impact on emissions and electrical conditions. It can take several hours for some rapper changes to begin showing impact on the precipitator performance.

It is desirable to have a slight buildup of dust on collecting plates. Dust depositing on the surface of the collecting plates will agglomerate with the dust already residing there. This reduces the potential for dust reentrainment during normal rapping. Generally, this dust layer should be less than $\frac{3}{16}$ inch thick and uniform across the surface of the panels.

If the dust layer is too thick, the potential exists for excessive amounts of dust to be dislodged during rapping. In addition, if the dust resistivity is high, the dust layer will create a voltage proportional to the resistivity of the dust. This will reduce performance of the unit.

The high-voltage system should not have a normal dust layer. It is desirable to keep the electrodes clean during operation. Dust depositing on the electrodes can create a voltage drop that will impair performance.

5.8.6 Insulator Cleaning

The high-voltage system is isolated from ground by support insulators. These insulators are exposed to process gas, which contains dust and moisture. Dust and moisture accumulating on the surface of insulators will cause them to track and carry current. This can result in loss of current necessary to charge dust and, in the extreme case, failure of the insulators.

In an electrostatic precipitator, there are insulators supporting the high-voltage system, stabilizing the lower high-voltage frames, and isolating the high-voltage rapping system. External to the process are insulators supporting the high-voltage bus and providing high-voltage termination from the transformer rectifier. All the insulators must be kept clean and free from carbon tracking.

5.8.7 Purge Heater and Ring Heater Systems

Most electrostatic precipitators operate under negative process pressure. As a result, air drawn into the penthouse or insulator compartment can cause condensation of moisture contained in the gas stream. The condensation results in accelerated corrosion and excessive sparking in the electrical field.

It is advisable to provide a blower filter heater arrangement that forces air into the insulator enclosure. This clean heated dry air will mix with the process gas without causing condensation.

If a purge heater system cannot be used, ring heaters installed around each support insulator will provide some protection.

It is essential that the purge heater or ring heater system be energized at least 4 hours before introducing process gas into the electrostatic precipitator.

5.8.8 Process Temperature

As indicated in the resistivity section, elevated gas temperature on a cold-side precipitator will result in degraded performance. As a result, it is critical to minimize process temperatures entering the cold-side unit.

This can be accomplished by monitoring soot-blowing programs and maintaining the heat transfer efficiency of the air heater.

In the case of a precipitator operating on the hot side of the resistivity curve, it is beneficial to maximize gas temperature. When operating this type of unit at reduced load, high-resistivity dust may build up on the collecting plate and electrodes. This will result in excess emission during load ramp up. To avoid this problem, an aggressive rapping program should be initiated at reduced loads.

5.8.9 Fuel Changes

As coal composition changes, the resistivity of dust created can increase. Increased dust resistivity may result in reduced electrostatic precipitator performance. To alleviate this problem, it is common to increase the moisture content of the flue gas when operating on the cold side of the resistivity curve.

Moisture content of the process gas can be increased by operating the steam soot blowers or by installing an evaporative gas conditioning system ahead of the precipitator. If alternative coals are on site that have more favorable resistivity, they can be blended with the difficult coal to produce better precipitator operation. In severe cases, it may be necessary to install a flue gas conditioning system that injects SO_3 into the gas stream.

6

Evaporative Coolers

Wayne T. Hartshorn

Hart Environmental, Inc., Lehighton, Pennsylvania

6.1 Device Type

Evaporative gas coolers use the controlled application of a liquid (usually water) to a hot gas stream to reduce that gas stream's temperature through the evaporation of that liquid. The liquid is often applied in the form of an air-atomized mist or fog.

6.2 Typical Applications and Uses

Evaporative coolers are designed to reduce a hot gas stream's temperature to a level suitable for further treatment. They are also used to "condition" the particulate before capture in another device.

When a gas stream requires treatment by a device that is sensitive to gas temperatures as well as gas humidity (such as a fabric filter collector), an evaporative gas cooler is often used to reduce the gas stream temperature to a tolerable level above the saturation temperature. Through the careful application of the liquid, the outlet temperature can be reduced, yet the bulk stream quality can be maintained safely above the water saturation temperature and/or acid dewpoint.

The evaporative gas cooler is sometimes also used ahead of devices such as electrostatic precipitators or spray dryers to temper or condition the gas stream before particulate separation or gas absorption onto a sorbent. For boiler applications, the addition of moisture often favorably reduces the resistivity of the fly ash.

Evaporative coolers are often used as the first stage of a gas cleaning system on hot gas applications such as thermal oxidizers, incinerators, furnaces, calciners, and kilns. Figure 6.1 shows an evaporative cooler (to the right) ahead of a pulse-type baghouse equipped with dry lime injection on a medical

FIGURE 6.1
Evaporative cooler on pulse type baghouse (Bundy Environmental Technology, Inc.).

waste incinerator. The evaporative cooler reduces the flue gas temperature to less than 500°F to protect the filter medium in the collector and to reduce the treated gas volume.

6.3 Primary Mechanisms Used

Evaporative coolers use the heat of vaporization of a liquid to extract heat from the gas stream and thereby reduce the mixture temperature.

The evaporation rate is dictated by the temperature and humidity differential between the desired outlet gas conditions and the given inlet gas quality. The droplet size produced by the evaporative cooling nozzles or spray system dictates the evaporation time and therefore the physical size of the evaporative cooler.

6.4 Design Basics

Over the years, much progress has been made in the further development and improvements of air pollution control (APC) devices, such as electrostatic precipitators (wet and dry), fabric filters (baghouses), scrubbers (wet and dry), as well as other types of collection equipment. However, far less attention has been given to the cooling and conditioning of hot-process gases before being treated in APC devices. Every APC device installed on a high-temperature application is affected in some way by the cooling technique used. Because of this effect, the area of cooling and conditioning becomes significant and indeed important when designing an overall gas handling or pollution control system.

Evaporative cooling can be applied to hot-process gases in many industries and applications. Some of those industries are ferrous and nonferrous metals, rock products, industrial and utility power, and incineration. When evaporative cooling systems are properly engineered, they can provide the most cost-effective method of dealing with increased heat loads from these sources.

6.4.1 Types of Gas Cooling

The three most used techniques for cooling hot-process gases are dilution cooling, convection/radiant cooling, and evaporative cooling. Figure 6.2 shows the effect of evaporative and dilution cooling on resulting gas volume when cooling to 400°F. When selecting an APC device to be installed downstream of the gas cooling system, it is important to note the lower gas volume that results using evaporative cooling versus dilution cooling.

Dilution cooling is the use of ambient air to dilute the total heat content of a hot gas stream so that its resulting temperature is lower, that is, fewer British thermal units (BTUs) per pound of gas.

Convection/radiant cooling implies the use of heat exchanger surface to exchange BTUs from the hot gas stream to a suitable receiver fluid, which is normally air, or water in the case of waste heat boilers. The receiving fluid may be forced across the heat exchanger surfaces by means of fans or pumps, or natural convection currents can be used as in the case of hairpin-type radiant coolers.

Evaporative gas cooling is the use of the heat of vaporization of water to absorb BTUs from the hot gas stream and thus reduces its temperature. Evaporative cooling systems can be either wet or dry, depending on the design and the process requirements.

6.4.2 Gas Conditioning

When we discuss evaporative gas cooling, it is commonly understood that the concept is used to cool hot gases. However, evaporative cooling technology does more than lower gas temperatures.

The term *gas conditioning* can refer to many processes, but the result is to affect the nature of the gas in some way beneficial to the APC device. The purpose may be to change the gas or dust electrical resistivity, dust surface

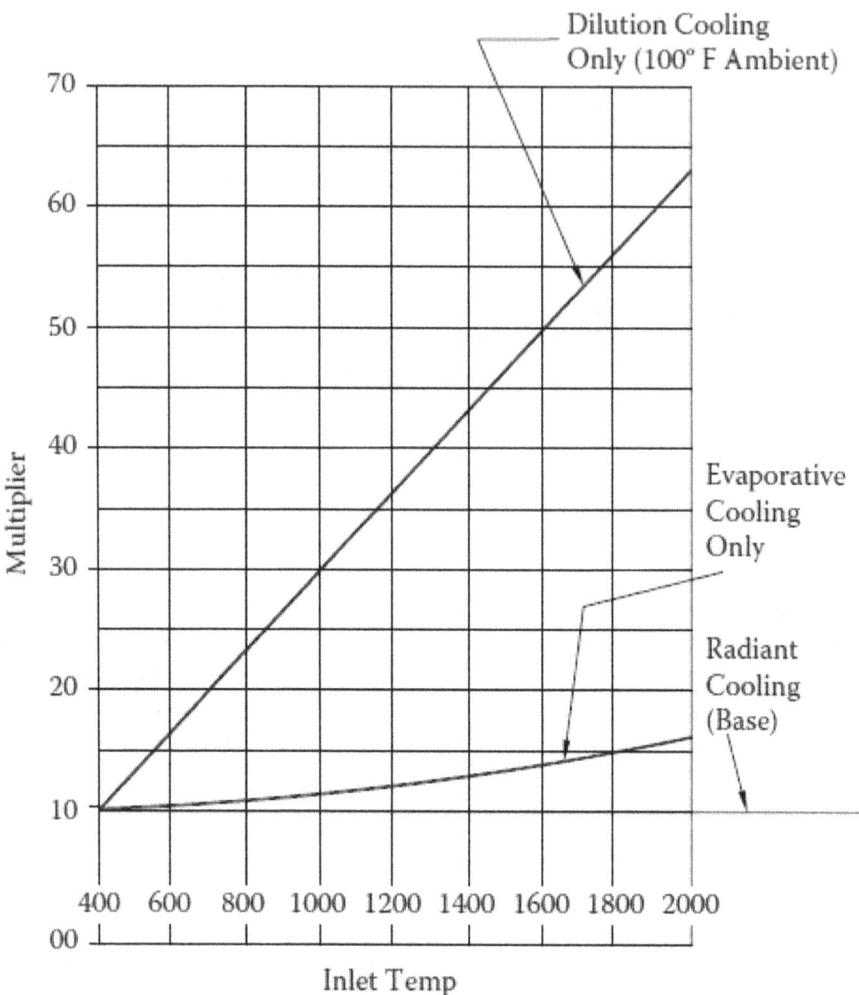

FIGURE 6.2
Effect of evaporative and dilution cooling (Hart Environmental, Inc.).

conditions, corrosion characteristics, odor, or many other functions. Gas conditioning is accomplished by the addition of water, acid, ammonia, or some other type of chemical. Figure 6.3 shows the effect of moisture added on fly-ash resistivity. Reducing the resistivity of fly ash can improve the performance of electrostatic precipitators.

The basic reason for cooling hot gases is to allow the gases to be collected by conventional APC devices, which have temperature limitations. There are some other reasons, however, which are somewhat less apparent and should be considered in the design of any APC system: to improve collection efficiency of the APC device, to reduce the size of the APC device and associated

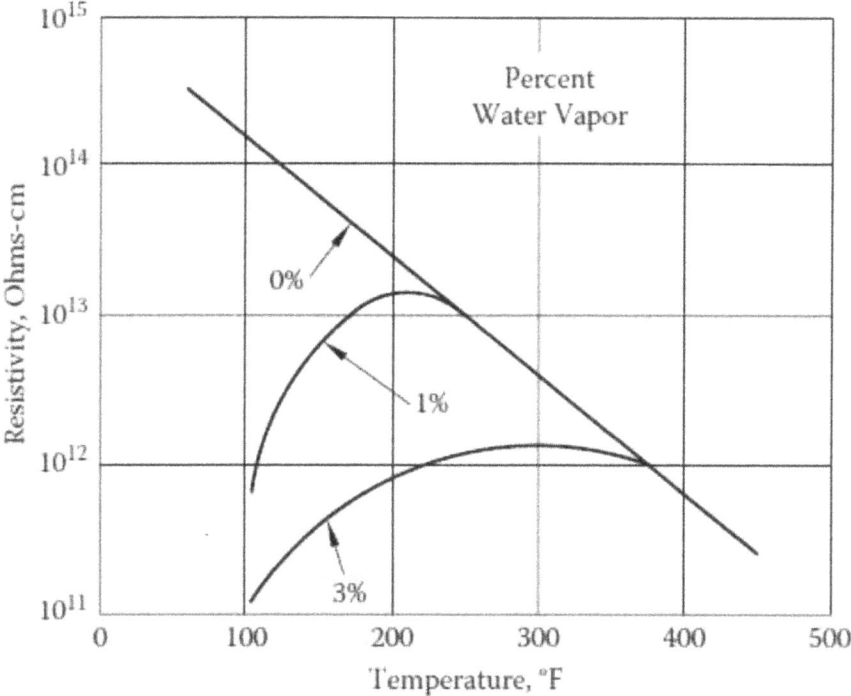

FIGURE 6.3
Effect of moisture on fly-ash resistivity (Hart Environmental, Inc.).

equipment, to reduce maintenance and thus downtime in the collection and related equipment, to increase production, and to improve reliability and service life of the APC device and components.

Both terms *evaporative cooling* and *evaporative conditioning* imply the injection of water into a hot gas stream. The purpose may be to reduce the gas volume by reducing gas temperature, to alter gas or dust properties by changing humidity, or to reduce temperature to allow less-expensive filter materials and/or materials of construction. Whatever the reason, the problems remain the same. Reviewing technologies around the world revealed two general groupings of problems with some types of technologies: problems of original design generally related to the sizing of the equipment, atomizing nozzle type and placement, and ability or inability to turn down; and mechanical and maintenance problems associated with the type of spray nozzles selected and the gas velocities in the systems.

Due to the history and problems associated with evaporative gas cooling and conditioning, efforts were put forth to improve the design and reliability of water spray systems on all industrial applications. Those efforts included a better understanding of why the systems were being used. In some cases, only cooling of the hot gases was required and it was not desirable to affect the properties of the suspended dust particles or gases. In other cases, field experience has shown that the real object of water sprays was to affect the electrical resistivity of the dust particles or gases, and cooling was simply a secondary function. In many industrial applications, the temperature and electrical resistivity level is critical when using a hot/dry electrostatic precipitator as the APC device.

When considering an evaporative gas cooling and conditioning system, one must bear in mind process requirements. Cooling equipment and components can be selected on the following basis: collection or APC device requirements, process outlet temperature, temperature cycles from the process, and the nature of the gas stream.

The first step is to select or determine the type of collection equipment or APC device that will be used for control of emissions. The properties of the emissions, the particulate loading, and the nature of the emissions will affect the type of APC device used. After the collection device has been selected or determined, the operating temperature must be determined. In the case of a dry electrostatic precipitator, electrical resistivity will be a factor. In the case of a baghouse (fabric filter), the type of filter material will be a critical factor; and the maximum temperature of the inlet gases will be a function of materials used and capabilities in the case of a scrubber or wet electrostatic precipitator.

After the collection equipment and the inlet operating temperature are known, the designer must consider the process outlet temperature to determine the amount of cooling required.

Another factor, which is important in the selection, design, and control of the evaporative gas cooling and conditioning system, is the gas temperature profile. Very constant profiles are easier to handle, but rapidly cycling temperatures are more difficult to handle and control. Knowing the process temperature profile will allow the designer to select the right control for the evaporative cooling and conditioning system. The control system or method of controlling rapidly changing temperatures can provide a very constant outlet temperature. It is extremely important to maintain a very constant outlet temperature from a cooling system to protect and maximize the efficiency of the APC device. A meticulously designed evaporative gas cooling and conditioning system can accomplish this requirement.

A rather serious consideration regarding the cooling system design is the effect of the cooling process on the chemical composition of the gas stream. There may be vaporous constituents, which condense at certain temperatures, through which cooling must be affected; and if a plastic phase is involved with that condensation process, extreme fouling or plugging of ductwork or other equipment may result unless cooling is effected rapidly. An intelligently designed and applied evaporative gas cooling and conditioning system can accomplish this rapid cooling or quenching.

6.4.3 Basic Sizing

There are three fundamental elements necessary for designing and selecting an evaporative hot-gas cooling and conditioning system.

1. A sound understanding of the dynamics of droplet evaporation under varying conditions.
2. Spray nozzles capable of producing extremely fine water or liquid droplets over a wide flow modulation range and with the ability of creating finer droplets with turndown.
3. A control system and overall systems view, which takes full advantage of the design data and modulation capability of the spray nozzles while recognizing and designing for the environment into which it is to be applied.

Evaporative cooling involves the use of fine water sprays to cool a hot gas stream. The cooling section is located between the heat source (furnace or process) and the APC device (dust collector equipment) and, in its simplest form, consists of a straight section of ductwork, or a chamber (usually cylindrical) with spray nozzles inserted through the walls. At times, the inlet gas temperatures exceed the temperature limits of ductwork or chamber steel. When this occurs, refractory-lined ductwork or chambers are used. When chambers are used, they are usually cylindrical and mounted vertically with the gas inlet transition at the top or bottom depending on the overall system design. In all cases, spray nozzles are positioned for maximum gas/water contact.

When using a cooling chamber, one must provide adequate residence time for droplet evaporation. The diameter of the cooling chamber is sized to limit gas velocity from 700 to 1200 feet per minute based on the average gas volume rate at the inlet and outlet.

Water usage calculations are made by performing an energy balance on the system. Using readily available enthalpy tables or specific heat data, the required flow rates of water are calculated for the expected hot gas flow rates.

A close estimate of the water usage requirements to cool hot gases can be made using results shown in Figure 6.4. The calculated data plotted in Figure 6.4 were obtained for the cooling of hot, dry air by water evaporation assuming constant specific heats for air and water vapor. Given the normal degree of fluctuations and uncertainty in the measured hot gas flow rates in industrial practice, the graph shown in Figure 6.4 yields quick information, good enough for most preliminary equipment sizing and design purposes.

More accurate predictions of water usage must be made using enthalpies for each gaseous constituent present in the hot-process gas stream.

One of the major shortcomings of traditional evaporative gas cooling and conditioning systems of old has been the lack of good quantitative data,

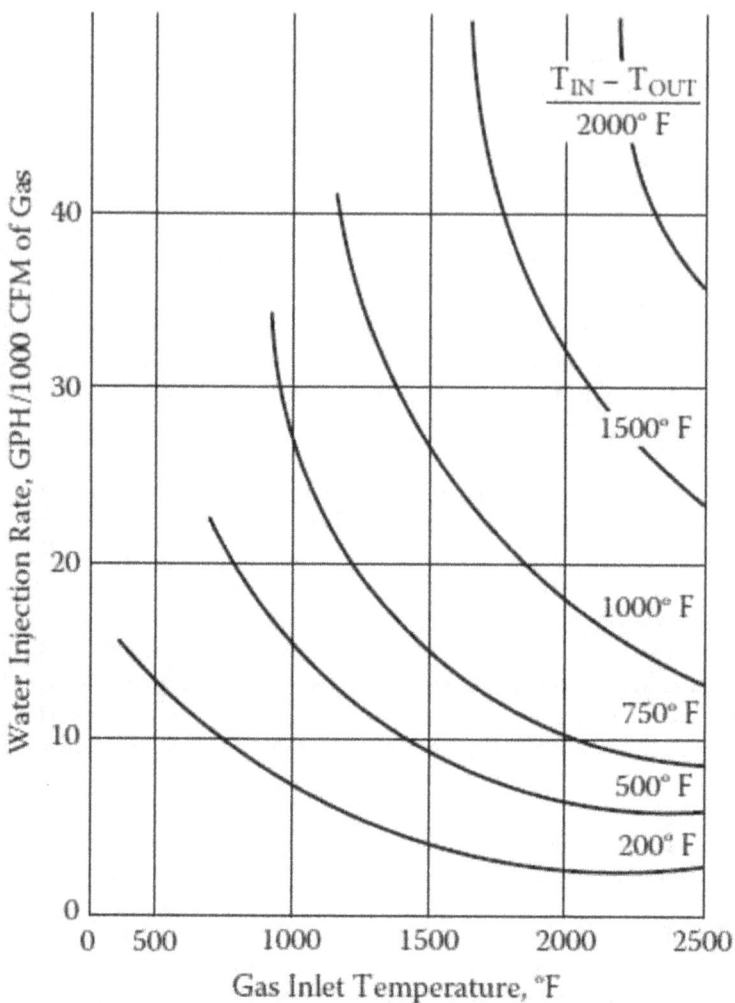

FIGURE 6.4
Water injection rates (Hart Environmental, Inc.).

which would allow accurate determinations of residence time. Accurate calculations for residence time determine the size requirements of either the cross-sectional area of ductwork or a cooling chamber and the length necessary to accomplish the total evaporation.

Many suppliers of spray towers base their designs either on data that were available for spray drying from Marshall but were applicable to only a small range of temperatures, or on other parameters that were based on other limited field experience or a good understanding of nozzle geometry, or both.

In 1972, an exhaustive computer analysis of evaporation rates was performed. The study analyzed evaporation rates of various droplet distributions with inlet temperatures ranging from 650°C to 1370°C under various conditions of inlet humidity and velocity.

The most significant findings of the study are as follows.

1. The largest droplet in each spray distribution required the longest time to evaporate. As simple and intuitive as that sounds, the importance was not previously recognized.

2. An excess of fine droplets in the presence of a few large ones increased evaporation time (t_e) by lowering the temperature (driving force) surrounding the larger droplets.

3. A determination of residence time cannot be made by a consideration of the largest droplet or the mean droplet diameter alone but must consider the entire droplet distribution. Effective droplet diameter (D_{eff}) for a given distribution is defined as the equivalent droplet of a perfectly homogeneous spray that would evaporate at the same time. The actual value of D_{eff} must be determined from the gas cooling supplier or from experienced gas cooling spray nozzle experience. The formula for determining effective droplet size is shown as follows.

$$D_{eff^2} = t_e \cdot f(T_s)^{10C_1 T_g C_2 T_i - C_3}$$

where:
  t_e = evaporation time
  D_{eff} = effective droplet diameter
  T_i = initial temperature
  T_s = saturation temperature
  C_1, C_2, C_3 = constants
  T_g = average temperature

The moisture content of the gases to be cooled cannot be neglected in the determination of evaporation time when the outlet temperature approaches T_s as in those cases $f(T_s)$ approaches 0 and t_e approaches infinity.

6.4.4 All-Important Atomization

Effective and reliable evaporative gas cooling and conditioning *must* begin with carefully selected and applied atomization. All atomizing nozzles are not created equal especially when using them for evaporative gas cooling and conditioning.

An atomizing nozzle used in this application (gas cooling and conditioning) should have the following characteristics.

1. Efficiently produce water droplets with small maximum droplet diameters and relatively uniform size distributions (minimum D_{eff}) at maximum flow rates.
2. It should have a wide-flow modulation characteristic while producing finer droplets with turndown. This is important because evaporation time increases as inlet temperature decreases.
3. It should be designed to minimize maintenance; that is, the nozzle materials of construction and design must be suited to operate and live in aggressive hot gas environments, utilize relatively large liquid ports to minimize internal pluggage, and be relatively self-cleaning to avoid external buildups of gas-laden dust, which would interfere with its atomizing characteristics. Figure 6.5 shows a photo of a heavy-duty gas cooling nozzle.

There are two types of atomizing nozzles that can satisfy the requirements for hot gas cooling and conditioning applications. These nozzles are first and foremost robust in construction, and second, capable of producing the kind

FIGURE 6.5
Heavy-duty atomizing nozzle designed for evaporative gas cooling (Hart Environmental, Inc.).

FIGURE 6.6
External mix nozzle (left); internal mix nozzle (center, right) (Hart Environmental, Inc.).

of droplet size distributions necessary for effective atomization. These nozzle designs are referred to as *dual-fluid atomizers*. This is where a liquid, usually water, and a compressible gas, usually compressed air, are pumped into the nozzle in combination to supply the liquid and energy for the required atomization. The two nozzles, both dual-fluid types, which will be discussed here, differ in geometry. One is referred to as an *external mix device* while the other nozzle is an *internal mix device*. See Figure 6.6 for a photograph of the two types of gas cooling atomizing nozzles.

External mixing is where the liquid, usually water, is introduced externally into the compressed air. Mixing the liquid externally with the accelerated air stream shatters the liquid into exceptionally fine droplets. In the internal mix device, the liquid and compressed air are mixed internally in a multiport fashion before exiting the nozzle outlet orifice.

The external nozzle generally uses more compressed air consumption but can produce turndown capabilities of as much as 20 to 1. The internal mix type nozzle will be more of an energy saver, but the turndown is lower at 10 to 1. Each nozzle design can be produced in many sizes, and both have their strengths and weaknesses. Depending on the process and system requirements, one nozzle type may have some advantages over the other. However, during the selection process, a systems analysis must be completed to decide which atomizing technology is best for a given application. Because both the external and internal mix nozzles do not rely on hydraulic energy to atomize, the liquid ports are relatively large, and wear does not affect performance within broad limits.

A significant advantage of the nozzles presented here is that controlling the ratio of energy to flow with turndown can control the size of the liquid droplets. This is an important aspect of the nozzle selection because it allows the cooling duct or chamber to be sized as a function of maximum temperature conditions without risk of low-end problems. Although other nozzles produce a similar degree of atomization, that is, extremely high-pressure hydraulic nozzles, these nozzles pose mechanical and operational problems, which preclude their general use. They use extremely small liquid ports that plug and wear, are limited in their maximum flow capability, do not offer adequate turndown ratios, and where droplets increase in size as the nozzles are turned down. Furthermore, these nozzles produce higher momentum directional sprays, which impinge on duct or chamber walls creating corrosion and dust buildup problems.

Although the data and spray nozzles provide the major technical components to this gas cooling and conditioning technology, they cannot stand alone. Each component of the overall system must be designed to survive the plant environment, to function through the full range of operating conditions, and to minimize maintenance. Some parameters that should be considered in this technology are as follows.

1. Gas inlet design: Gas flow through the inlet section of a cooling chamber or into a duct section where the spray nozzles are located must be straight to avoid washing of the walls. The use of internal distribution devices should be avoided.

2. Gas velocity: Gas flow direction through the duct or chamber is maintained using relatively high velocity, minimizing the potential of wall buildups.

3. Controls: Processes modulate on a continuous curve, not as a step function. Controls must modulate on the same curve and must be as rapid as the process to ensure that exactly the correct quantity of water is injected at any given time. Depending on several variables such as inlet and outlet temperatures, fluctuations in process conditions, and flexibility and adjustability requirements, the scope of the control can vary. Control schemes will vary from single-loop feedback control to feed-forward/cascading-type control. Some applications may be able to use pressures of both fluids to control flows, while other difficult gas streams require more powerful controls of actual flow measurements and specific algorithms for more precise control. The best control philosophy for a specific application must be determined during the review of the specifications and process conditions. Figure 6.7 shows a typical flow control scheme.

4. Redundancy: In certain critical areas, there is a need to provide redundant equipment to minimize downtime, which affects production. Utilities and measuring devices are generally the most vulnerable components and require the greatest attention.

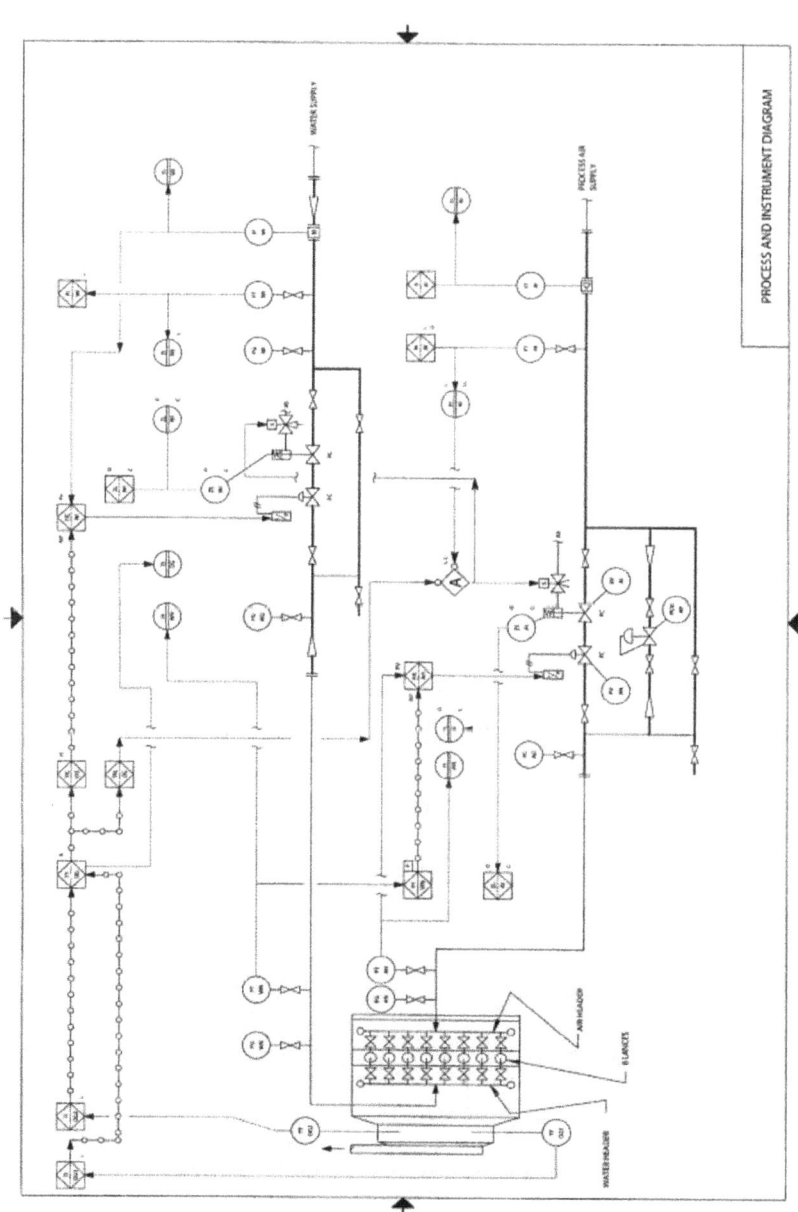

FIGURE 6.7
A typical process and instrumentation diagram for flow control (Hart Environmental, Inc.).

5. System layout: A good system review and layout will serve to minimize installation costs and maintenance by packaging related materials close to each other, minimizing ductwork and structural steel,
and facilitating access.

The system considerations listed previously are universal and do not apply
solely to evaporative gas cooling and conditioning systems. Evaporative gas
cooling and conditioning systems, whether duct cooling or cooling chambers, have a unique reputation for misapplication of the principle. They
typically do not get the attention they deserve compared with the selection
process of APC devices. Evaporative gas cooling and conditioning systems
are an important and integral part of the gas handling system. They should
be coordinated and thoroughly thought out as a key element to the overall
performance of any APC system scheme.

What does an evaporative gas cooling and conditioning system look like?
Figure 6.8 shows an evaporative cooling tower in use ahead of a baghouse

FIGURE 6.8
Evaporative cooler on aluminum plant (Hart Environmental, Inc.).

at an aluminum plant. The evaporative cooler is located at the center of the frame and the baghouse is to the left of center.

6.4.5 Case History Example

One of the earliest applications of improved dual-fluid atomizing nozzles to evaporative hot-gas cooling occurred at a mining company that processes copper ore into a refined product.

The roasting operation resulted in an 1100°F gas, which had as its major contaminating constituent sulfur dioxide and combustible dust particles, which would ignite if the cooling sprays were turned off. Because the dry electrostatic precipitator that was installed at that time could not tolerate temperatures above 800°F, an evaporative gas cooling tower was used to reduce the gas temperature to the desired level and to provide additional moisture for subsequent conversion of sulfur dioxide to sulfuric acid.

The original installation used 10 high-pressure, single-fluid, 350-psig spray nozzles in a vertical cooling chamber 31 feet high by 11 feet in diameter. The liquid droplets produced covered a wide size range, and their velocities were excessive due to the high atomizing liquid pressure. Approximately 15,000 cubic ft/min of hot gas with a velocity of 2.6 ft/s did not provide enough residence time for complete droplet evaporation due to the high velocities of the larger droplets produced by the high-pressure spray nozzles. The large droplets remained in the cooling chamber due to incomplete evaporation or impaction and run-off down the chamber internal walls together with entrained dust buildup, fouling sludge deposits in the collector's hoppers. Frequent shutdowns of the system were required to permit cleaning out the sludge buildup.

The changeover to correct and/or improve the evaporative cooling operation was to three dual-fluid external mix nozzles. The dual-fluid atomizers were adequate to replace 10 high-pressure nozzles that were originally used. The retrofitted dual-fluid nozzles operated at 60-psig compressed air and 58-psig water pressure. The resulting spray of approximately 3.5 gallons per minute (GPM) per nozzle produced droplets that minimized the unevaporated water fallout and sludge buildup. The unscheduled shutdowns due to problems associated with the original high-pressure spray nozzles were eliminated.

6.4.6 Cost Considerations

Experience over the years has indicated that the installed cost of a complete evaporative gas cooling system generally runs about 10% of the complete APC or gas handling system price. Because of this low cost, the evaporative gas cooling system has received much less attention than the APC devices. However, the design, selection, and performance of the evaporative gas cooling system can have a major impact on the success of the overall gas handling or gas cleaning system. This performance can complement the emission as well as the maintenance of the plant's operation.

A complete evaporative gas cooling system will consist of a cooling chamber or duct, the spray nozzles and supporting lance assemblies, air and liquid valve rack trains, compressed air, pumping station, and the necessary piping and wiring to connect the components.

The major consideration and operating cost for a dual-fluid nozzle system will be in the compressed air usage. Compressed air is the second fluid necessary to produce the most efficient atomization and droplet sizes for effective evaporative gas cooling. Depending on the type of nozzle (internal or external mix) applied, the compressed air requirement can be from 4 to 10 standard cubic feet per minute (scfm) per GPM of cooling liquid required. For estimating purposes, it will take about one horsepower of compressed air to produce about 5 scfm.

Evaporative coolers can be quite large because adequate time must be provided to allow the atomized sprays to dry to completion. Figure 6.9 shows a

FIGURE 6.9
Evaporative cooler on dry process cement kiln (Hart Environmental, Inc.).

large evaporative cooler on a dry process cement kiln in use ahead of a dry electrostatic precipitator.

6.5 Operating Suggestions

Many times, the fate of the downstream equipment is in the hands of the evaporative cooling system. On the high-temperature side, the bags in a baghouse may not be able to tolerate temperatures much more than the desired operating temperature. On the low temperature side, excessive cooling can bring the gas stream temperature and humidity to at or near the acid dewpoint. The evaporative cooler should be meticulously designed and maintained.

Make certain that the worst condition of the liquid you are using is made known to the evaporative cooling system vendor. Under hard water conditions, it is often wise to use softened water to reduce nozzle scaling and/or plugging. Filtered water is an absolute minimum requirement.

Allow in your design space to pull and spray lances or headers. Stage the spray systems if possible and allow for back-up spray assemblies to be used. The control system should also be anticipatory; that is, monitor trends in gas temperature versus the evaporative system response. If the controller senses that it cannot keep up with the evaporative demand, suitable alarms, or even shutdown should be activated.

Any feed pumps and compressors should be redundant if possible if the application is extremely heat sensitive (say, the source is above 1200°F).

Likewise, the cooler outlet temperature thermistor or thermocouple should be redundant to make certain that this important signal is clean and constant.

Spare nozzles (as a minimum) and spare lance assemblies should be purchased and kept in stock so that the evaporative cooler can be maintained at peak operating performance.

Do not skimp on evaporative cooler vessel size. An excessively small vessel can allow mist carryover to the downstream equipment and cause corrosive damage or, in the case of a baghouse, bag blinding.

A professionally designed evaporative cooler will temper a hot gas stream reliably, day after day, and help make any downstream equipment perform at its best.

7

Fabric Filter Collectors

Deny Claffey and Michael Claffey
Allied Mechanical, Las Vegas, Nevada

Jerry Childress
McGill AirClean, Columbus, Ohio

7.1 Device Type

Fabric filter collectors, or baghouses, separate particulate from gas stream by causing the particulate to pass through a filtering medium, a layer of previously collected (or purposely deposited) particulate, or both. The gas-borne particulate is intercepted by the fibers of the filtering medium, by the particulate already present on the medium surface, or both. To prevent excessive pressure drop as the particulate accumulates, these devices use various mechanisms to disengage the particulate from the medium.

7.2 Typical Applications and Uses

There are three basic dust collector applications. *Nuisance* or *fugitive* dust collection from the venting of conveyors, transfer points, packing stations and so on—this dust is often sent to waste. Next is *product collection* venting of classifiers, crushers, storage bins, air (pneumatic) conveying systems, mills, and flash dryers. This dust is often recovered because it has value. Last is *process gas filtration* venting of spray dryers, kilns, power boilers, reactors, and so on. The collected solids may or may not be returned to the process. This dust may or may not be worth recovering but must be controlled for environmental or workplace health reasons.

Fabric filter collectors are also currently used for gas absorption applications wherein the fabric filter collector is preceded by a spray dryer, dry Venturi, ductwork injection system, or the bags are precoated with an adsorbent or absorbent. Sodium bicarbonate precoat, for example, has been used

FIGURE 7.1
Baghouse with preconditioner (Bundy Environmental Technology, Inc.).

to remove gaseous SO_2 from power boiler exhaust gases. A precoat of lime or spray-dried slurry of lime has also been used on many applications to simultaneously remove particulate and acidic gases. When toxic dioxins are present, some applications use activated carbon as part of the precoat.

Figure 7.1 shows a baghouse preceded by an evaporative cooler on a cupola operation. The hot gases enter from the bypass stack at the left and proceed to the downward firing cooler/conditioner. An absorbent is injected in the vertical cylindrical tower at the center of the picture. Toward the right is the baghouse in which the absorbent and process particulate is collected. The stack is on the right.

In contrast in size and complexity, the small dust collector in Figure 7.2 collects dust from problem sources and deposits it directly into a drum.

Fabric filter collectors are generally *not* used where the particulate (dust) is combustible or where the product is to be sent back to the process and wetted. For the latter, it is often easier to simply use a wet scrubber for collection. In that manner, the product is prepared to be returned to the process. Fabric filter collectors are also avoided if glowing embers or other such

FIGURE 7.2
Nuisance dust collector with drum (American Air Filter).

damaging carryover exists that could damage the collecting media or cause a fire. In some cases, a suitably designed cyclone collector is used to protect the baghouse.

More recent applications of fabric filter collectors include the collection of both particulate and acid gases. To accomplish this, a spray dryer, dry Venturi, or other dry powder injection mechanism is added ahead of the fabric filter collector. The dry powder (typically hydrated lime or sodium bicarbonate) is mixed with the gas stream and is pneumatically conveyed to the fabric filter collector. In the fabric filter collector, the residual powder adheres temporarily to the gag medium and provides an adsorptive surface. The designer must allow for the gas flow resistance of this *filter cake* in sizing the bag surface area and the selection of the prime mover (usually a fan). The collector itself may use pulse jet cleaning or shaker cleaning, but in each

FIGURE 7.3
Spray dryer and fabric filter for acid gas control (photo courtesy of McGill AirClean).

case the bag material selection must be compatible with the filter cake characteristics since some of the powder is purposely held on the bag surface to enhance the absorption. Also, to obtain adequate acid gas removal efficiency (above 85%–90%), the gas stream must be operated at close to the acid dewpoint (usually within 50°F of the dewpoint) since the gas stream water vapor content contributes to the collection of acid species. Figure 7.3 shows a spray dryer followed by an eight-compartment fabric filter collector used to control SO_2, HCl, and particulate from a manure-fired bubbling fluidized bed boiler.

Another application carries the above one step further. For the control of mercury (Hg) emissions particularly from thermal waste disposal equipment or coal-fired boilers, powdered activated carbon (PAC) is also dry injected. The activated carbon is used to adsorb the mercury and retain the mercury for subsequent disposal. In one installation, the PAC is injected into two spray dryers (as shown in Figure 7.4), which are then followed by fabric filter collectors. This installation was applied to two rotary kiln municipal solid waste incinerators.

FIGURE 7.4
Mercury control using PAC (photo courtesy of McGill AirClean).

7.3 Operating Principles

Fabric filter collectors' function by filtering or screening particulate from the gas stream that carries that particulate. To understand this better, first, a little bit of history.

Dry dust collectors have evolved through the years from very primitive basic designs to a relatively sophisticated series of machines. Initially, when air pollution control regulations did not exist, collectors were only required to catch some of the particulate coming off a process. For example, at one time a drop-out box (settling chamber) could in some cases meet the collection criteria. The dry cyclone was, for a time, the ultimate in collection machinery.

These first dust collectors were simple mechanical machines. The drop-out box (settling chamber) took a moving air stream, including dusty particulate, and slowed it down to a point where the particulate dropped out due to its own gravity. The slower the air velocity, the heavier the particulate, and the better the separation. The biggest box allowing for the lowest air velocity and longest retention time was the best. In the real world, the drop-out box was then and still is well suited to separate lighter floating products from heavy particulate. The lesson here is that gravity and carrying air velocity are still particularly important issues to consider in any dust collector, but they have their limitations.

As mentioned in Chapter 4, "Dry Cyclone Collectors," the dry cyclone uses gravity and centrifugal force to spin the dust out of the air. Cyclone designs can be very sophisticated, and they can be extremely efficient solid-separation devices and classifiers. Cyclones at one time could separate enough dust from processes to be considered an air pollution control device. Centrifugal force alone was not enough. As time went by and air quality standards became more stringent, a fabric filter collector became the primary device to use to meet air quality standards. In applications with high particulate loadings or when processing stringy floating type products, a cyclone makes an excellent scalper or precleaner for a fabric filter. A cylindrical fabric filter with large annular space between filters and shell setup with a high tangential cyclone type inlet is an excellent heavy-duty collector/receiver.

Fabric filters are devices that use some type of permeable fabric to screen the particulate from moving air. This fabric or material is often called the *filtering medium* or simply, *medium*. The first fabric filters were panel-type designs somewhat like a home hot-air furnace filter, but their time was short lived because they could not self-clean. As they plugged or blinded, they were changed manually, discarded, and replaced with new filters. The next step was to develop a machine with a fabric filter that could clean itself. The first devices used tubular fabric socks arranged in rows in a matrix enclosed in a housing with a hopper. There were basically two types: the shaker and the reverse air type. The pulse jet collector followed. All these devices used tubular socks of media arranged inside a housing above a hopper to catch the particulate as it was cleaned off the vertically mounted bags or tubes. These baghouses incorporate a tube sheet that holds the bag filters in place. The tube sheet also separates the collector into a clean side and a dirty side arrangement. The clean air side is called the *clean air plenum* (CAP) and the dirty air side is *dirty air plenum* (DAP). The hopper is located below the DAP, so gravity helps drop the dust into the hopper. The conventional dust collector is designed to get rid of the dust in the hopper immediately as it is generated. A filter receiver-type collector has an oversized hopper designed to hold dust/particulate for some time while the collector is still processing the dirty air stream.

7.4 Primary Mechanisms Used

Fabric filter collectors primarily use sieving (a combination of impaction and interception) as the collecting mechanism. The combined porosity of the medium and any previously accumulated particulate serve to produce small pores through which the new particulate must attempt to pass. This filtering or sieving action relies on the fact that the net opening at any given time is smaller than the particulate. Because the particle is bigger than the opening, it cannot pass through. After collection on the medium surface or in the dust cake, various mechanisms are used to remove the particulate from the medium. After that, the particulate settles by gravity in the device's housing.

7.5 Design Basics

The factors that affect sizing and performance of a collector are the material (dust) itself, the temperature effect on the air, gas, product, fineness of the material (fume being an example), dust, and particulate loading in grains per cubic foot. These factors determine the type of collector selected, the housing construction required, inlet locations, fabric medium selection, and dust discharge parameters. Dust collector manufacturers distribute application data inquiry forms that provide the answers to questions needed to specify the correct collector design and arrangement for a given application. For example, it is important to know if the dust is explosive, statically charged, hygroscopic light, heavy, fine, wet, sticky, and so on. Do we need insulation, hopper heaters, and special equipment for discharging dust? Is the collector located inside or outside? Does the exhaust air go back to plant or outside? These are just a few serious questions meant to indicate just how important it is for us to know the details before specifying any collector

After analyzing these parameters, the designer can then choose from among a wide variety of fabric filter collectors to solve the emissions problem. The most basic type is the shaker collector, named after its use of a shaking mechanism to dislodge accumulated particulate.

The shaker collector has tubular socks of a woven medium suspended by a strap on the top of the bag connected to a mechanical shaking arm. No cages are required to hold the bags open, and the lower end of the bag socks are clamped to the tube sheet located directly above the hopper. The dirty air enters the unit in the hopper section and is forced to go upward inside the socks. When the socks get plugged (blinded), the differential pressure goes up. This creates an electrical signal that shuts off the fan or closes a damper and shuts off the air flow into the collector. The shaker mechanism then shakes the filter socks for an adjustable period, dislodging the dust cake and allowing it to fall back down into the hopper. Shakers use a light woven fabric medium designed to be very flexible. After a time, the shaking stops, the damper opens, and air flows through the collector. The problem with the shaker is that it cannot operate continuously because the process air and ventilation system must be shut down for it to clean. To achieve continuous operation, compartmentalized shaker units with some modules operating on-line cleaning process air and with some modules off-line cleaning filters are required. Also, the light-woven, flexible filter medium is not particularly efficient at removing the dust from the air, making the shaker suspect as an air pollution control device. The shaker is considered a low-energy intermittent use collector. The filter medium does not get worked very forcefully during cleaning, which can be an asset relative to filter life in high heat or corrosive applications.

The reverse air collector is built in numerous configurations. It is a moderate energy device. Generally, it uses a caged needled fabric tubular medium, making it a rather good choice for air pollution control applications. The reverse air cleaning principle is to use an extra air mover for cleaning filters. This extra air mover produces a higher pressure than the air flowing through the collector; hence, a flow of air through the cage and medium from the clean side of the filter dislodges the particulate from the dirty side, allowing it to fall into the hopper. The frequency and duration of the cleaning cycle is much the same as the shaker type. This reverse air flow is usually better at cleaning than gently shaking the filter bag. The time of the cleaning cycle is much the same as the shaker. Again, this is particularly true when the collector is setup in modular fashion with some sections of the collector on-line cleaning process air and some sections off-line cleaning filters. Cleaning the filters off-line is easy because there is no process air pressure holding particulate on the filter bag surface. The only real problem with reverse air collection is that controlling the air, on and off, during cleaning cycles on modular arrangements is complex and costly. Figure 7.5 shows an industrial reverse air collector. The moving arm in the center of the vessel applies a reverse pulse of air to individual tube rows. Other reverse air collectors break the housing into compartments using isolation valves. Using blowers, the air flow through the compartment being cleaned can be reversed, thereby cleaning the medium.

Some reverse air collectors are built with tube sheets low, directly above the hopper, with dirty air flowing upward inside uncaged bags, and with the tube sheet high, under the CAP, with dirty air flowing to the outside of caged bags. Reverse air collectors are also built in a tall cylindrical form configuration as in Figure 7.5. Typically, operating on-line, a continuously revolving arm blowing the higher cleaning air pressure down inside the filters is used, as in our previous example. The solid product falls between the bags into the hopper. The round unit with the single revolving cleaning arm is a single module, cleaning a few filters at a time on-line, making it a stand-alone collector. This is especially true when the reverse air cleaning fan is located within the collector. Some models require an external fan or blower for cleaning energy, which adds to complexity, cost, space, and moist air cleaning potentials.

An inherent problem with round collectors is that they do not use filter space well. Many more filters can be in a square or rectangular configuration. This becomes increasingly important in large installations in the

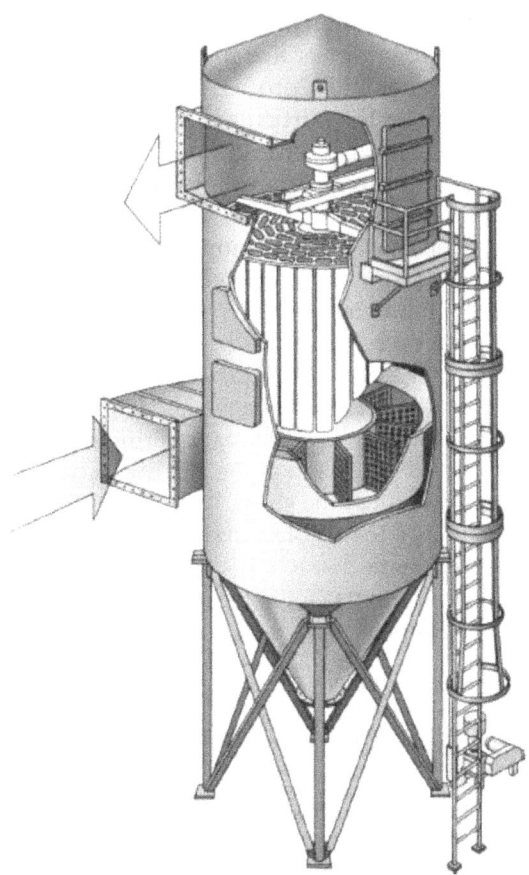

FIGURE 7.5
Reverse air collector (Donaldson Company, Inc.).

space-saving sense. Also, the tall form, cylindrical design does not lend itself to the architectural aesthetics of the modern low-profile industrial park. However, all in all, the reverse air does excel in some applications, especially grain, wood, paper, and other floating particulate. The cleaning cycle off-line is long enough to free the dust from the filter for an appreciable time so it can drop into the hopper. The model with a low tube sheet with uncaged filter bags is a good choice for heavy loadings in hot lime, cement, and kiln processing applications as the cleaning energy is not too intense to break filters down. Also, the mineral product is heavy enough to drop out of the bags and gravitate to the hopper.

The pulse jet collector is a high-energy cleaner as it uses high-pressure air blown down inside caged filter bags in bursts of 20–80 ms. The pulse jet uses filter bags with cages that are suspended from tube sheets between the DAP and CAP. Needled felt filters are used for high-cleaning-efficiency style, making it a good air pollution control device. This high-pressure air is typically directed through a Venturi to increase air volume, raises the air pressure inside the filter over the process air flowing through the collector, and the shock wave blasts the particulate off the filter bag, where it drops into the hopper. The pulse jet can be round, square, rectangular, short, tall, large, or small. It can be modified easily for trough, pyramid hoppers, high or low inlets, and walk-in or trapdoor CAPs, allowing for service in a clean-air atmosphere. It uses common factory-compressed air for cleaning instead of an extra fan or positive displacement blower. Some problems associated with pulse jets are that the high energy imparted to the filter breaks filter media down, particularly in high heat and/or chemical corrosive atmospheres. Also, the location of the Venturi is important with respect to the tube sheet. With the Venturi located in the filter bag itself, a negative air pressure exists above the Venturi lip down in the bag area, creating a suction pressure rather than positive air pressure at the top of the bag during cleaning. This leads to buildup of product under the tube sheet. It also takes the filter area of an 8-ft bag and effectively turns it to that of a 7-ft bag. A Venturi above the tube sheet eliminates this phenomenon.

The isometric view of a pulse jet collector is shown in Figure 7.6. In this unit, the gas inlet plenum is shown to the lower left and the cleaned gas

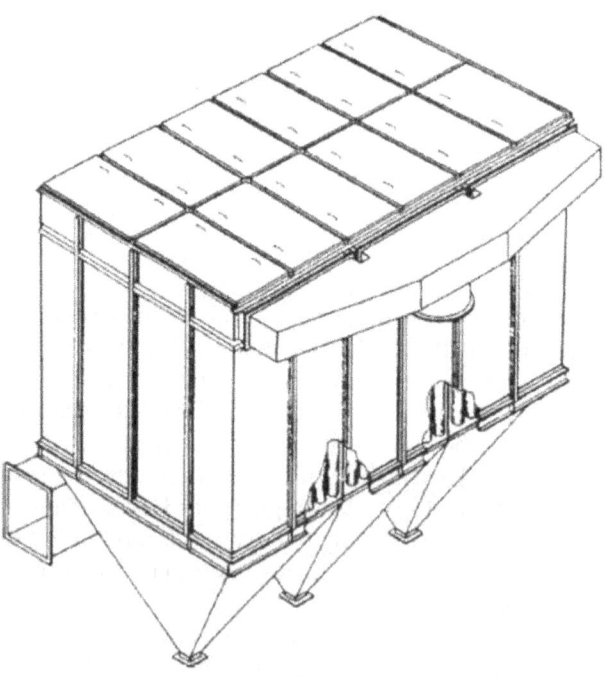

FIGURE 7.6
Pulse jet baghouse (Bionomic Industries, Inc.).

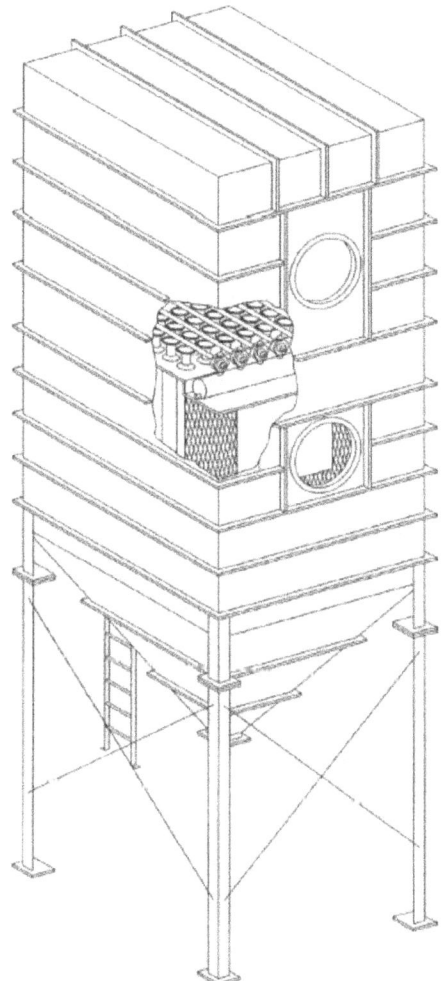

FIGURE 7.7
Pulse jet collector with high gas inlet (Steelcraft Corp.).

outlet is at the upper right, as part of a discharge plenum. The cutaway shows the bags arranged in rows in the collector. The bag access is through the top of this design. The rectangular sections at the top of the collector are doors that are removed for bag and Venturi access.

Pulse jet collectors can be configured in a variety of ways. In some cases, the gas inlet must be located up high. Figure 7.7 is of a pulse jet collector designed with a high gas entry inlet. It is also equipped with a "walk-in" type CAP (the chamber located above the Venturis). Figure 7.8 shows a similar collector but equipped with a low-level gas inlet.

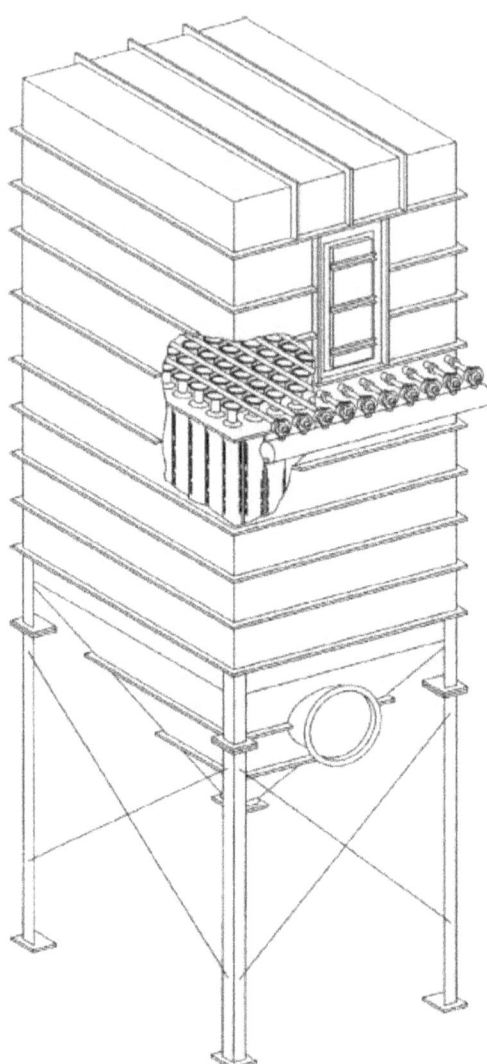

FIGURE 7.8
Low gas inlet pulse jet collector (Steelcraft Corp.).

Pulse jets can blast dirty, tacky product off the bag. If the particulate is moderately heavy or in clumps, it will drop into the hopper. If it is light or floats easily, it can get pulled right back onto the bag immediately after the short-duration cleaning pulse. Pulse jet self-cleaning cylindrical cartridge dust collectors use nominally 6–14-inch diameter × 26-inch-long pleated filters. Typical designs are shown in Figures 7.9 and 7.10. They were originally thought of as clean air filters because the filter design and cellulose media

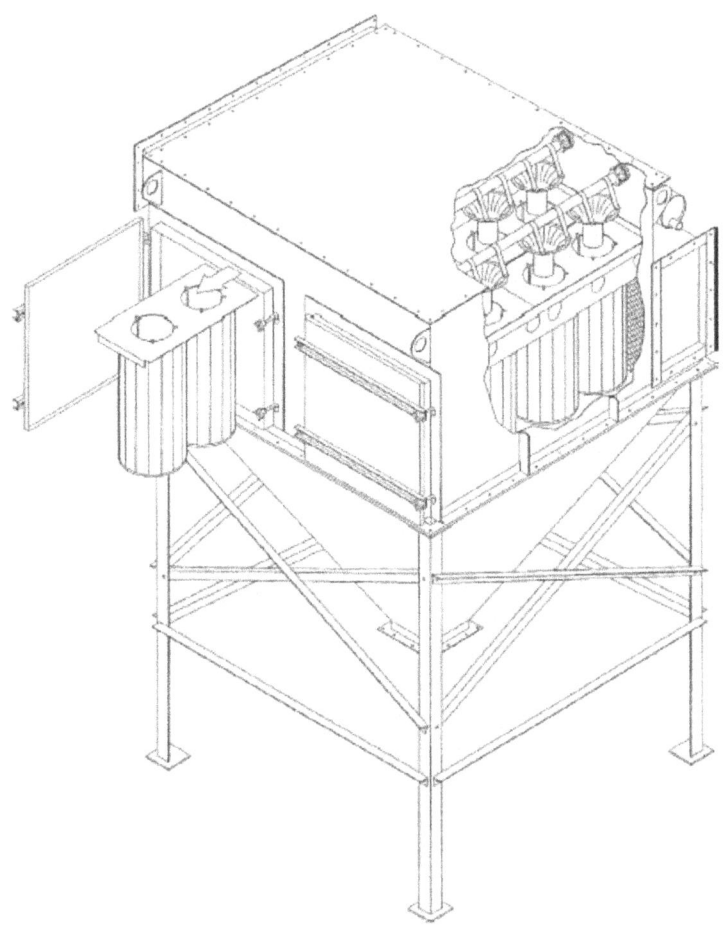

FIGURE 7.9
Cartridge collector (Steelcraft Corp.).

type provided high cleaning efficiency. They were and still are used to clean ambient air or as final filters (after filters) following heavy-duty conventional fabric filter grade collectors. The pleats provided much more filter area than a round 4- or 6-inch diameter tubular bag. The filters less cages were short and easy to handle. The collector holding them could be compact. Filter service could be done in clean air outside the collector on the side of the unit. The problem was initially as cartridge units started to be sold as true front-line industrial collectors, the tight pleats would plug up due to heavy dust loading and blind the filters prematurely. To solve this problem, the perforated metal around the periphery of the filters was removed and pleat spacing was opened so dust could be blown out of the pleats easier and off the filter. Heavy-duty spun bond polyester media became popular. Filters were made with filter bag

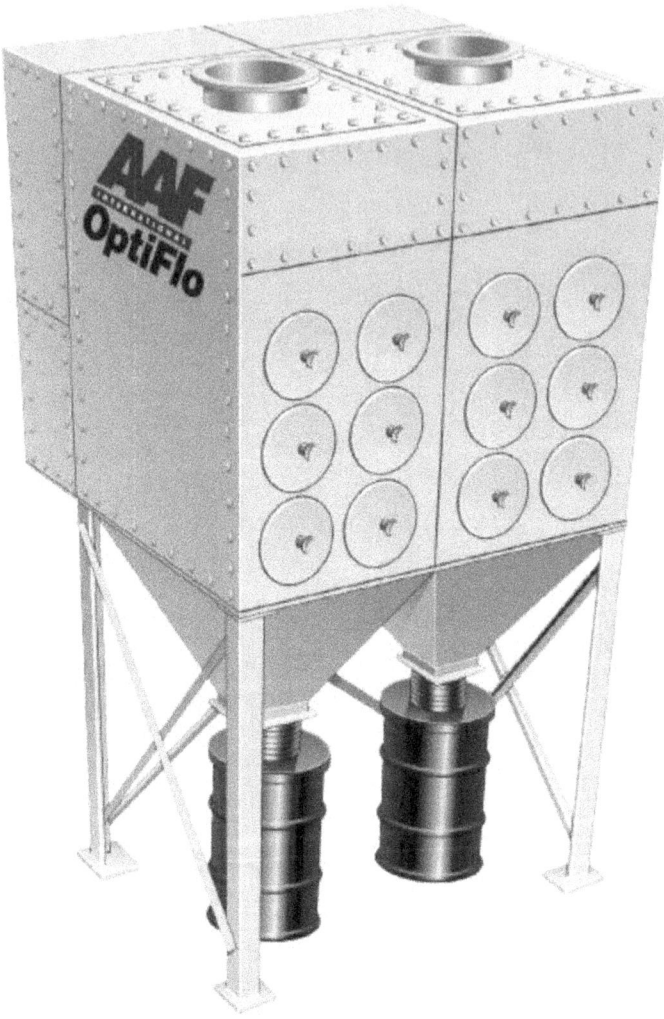

FIGURE 7.10
Side-access cartridge collector (American Air Filter).

geometry allowing for replacement of round filter bags in other type collectors with pleated filters (more area) in the 4–6-inch diameter range. Currently many styles of self-cleaning pleated filters are used in industrial processing. They are compact, service easily, and can tolerate moderate loadings at high levels of cleaning efficiency. They use compressed air for cleaning energy like pulse jet baghouses. Although they are still not the best for heavy loadings and aggressive dusts, pleated filters continue to gain in the industrial marketplace. The fact is, nothing cleans easier than a smooth, round shape.

There are many types and versions of dust collectors within the various types. This is because there are myriad applications, and certain designs are best suited for certain applications. In selecting a collector for a given job, it is critical to understand the details of the process completely. It is also critical to understand how the collector works in detail so a match can be made.

Basically, the best dust collector for the job will require the least overall cleaning energy and cleaning cycles to perform. It will operate at low pressure differential over the filters, holding fan energy down, and will provide long efficient filter life and infrequent service.

This tells us that when the dirty air enters the collector, the dust/particulate should take the shortest path to the hopper discharge and out. The filters should see only the lighter particulate/dusts that will build up a permeable filter cake to be cleaned off occasionally.

The prime considerations in collector design are inlet location and velocity and direction of dirty air flow inside the collector. For example, if the inlet is located below the filters, especially in a pyramid or conical hopper, all the air must go upward, directly impinging particulate into the filters. As the air/dust flows up between the filters, the air velocity (rising) increases, carrying the particulate up again and again into the filters. The dust has a hard time getting down past the inlet blast of air into and out of the hopper. On the other hand, if the dirty air inlet is located near the top of the filters, the dirty air flow must go downward directly toward the hopper or at worst horizontally onto the side of the filters. When filters need cleaning, the dust/particulate cake simply drops off into a quiet hopper, reducing any potential for air pushing the dust upward back onto the filter medium.

Sizing fabric filters starts with an air-to-cloth ratio that field experience has shown will work on a certain application. The air (cubic feet per minute) to cloth (medium area) calculation gives us the face or *impact velocity* of dirty air as it hits the filter medium. Let us assume we have a ventilation process requiring 7200 acfm and the suggested ratio is 6/1; therefore, 7200/6 = 1200 ft cloth required in the dust collector (nominally), and 7200/1200 = 6 cfm/ft² or 6 ft/min face velocity. As you can see, this provides us with a relative value for the volume and velocity of dirt and air flowing through the surface of the medium. The higher the gas velocity, the harder it is to push the dust off because you are pushing the dust back into the oncoming gas stream.

When using a compartmentalized off-line cleaning system, air-to-cloth ratio is a much less important factor as no process air is flowing into the filters. Cleaning off-line is quite easy at any air-to-cloth ratio.

Let us assume, again, that we are comparing two collectors, both processing 7200 cfm. The ratio being considered is nominally 6/1, meaning we need about 1200 ft² of filter media. One collector, the tall unit, needs 60 filters/cages at 6.2-inch diameter × 12 ft long to get approximately 1200 ft media. The filters are located on an 8-inch center grid pattern. The housing in the plan is 33.2 ft²; the filters in the plan, 11.76 ft². The open area between the filters is 33.2 − 11.76 = 21.44. So, 7200 cfm/21.44 = 336 ft/min velocity. The other collector, the short one, needs 90 filters/cages × 8 ft long each. With all the other parameters and geometry, the same, the velocity between filters is only 233 ft/min. About 30% lower! The tall filter will be cheaper because it will have fewer filters, cleaning valves, and a smaller housing, but the fact is it will not perform as well as the shorter, fatter unit.

One way to determine acceptable can velocity as it relates to air-to-cloth ratio collector performance is to use an industry rule of thumb for maximum allowable rising velocity on particulate:

120 ft/min max for up to 10 lb. cu/ft product

240 ft/min max for up to 20 lb. cu/ft product

300 ft/min max for up to 30 lb. cu/ft product

360 ft/min max for up to 50 lb. cu/ft product

400 ft/min max for up to 70 lb. cu/ft product

Using lower velocity is always best. Products that float, like ultrafine light dust, bees' wings, feathers, and fiberglass fines, all need special consideration. Use collectors designed for that service. What we are doing here is comparing the terminal settling velocity of the dust particle in a relative sense to the velocity of the air between the filters. Four hundred mesh soft wood flour at 8 pcf is much harder to drop out in a hopper than 30 mesh silica sand at 75 pcf. Grain husks, paper trim, and fiber from buffing wheels act differently than 94 pcf Portland cement. Selecting or specifying a collector is really a matter of common sense and the experience of the user or manufacturer. In some cases, like dry SO_2 removal, we want a coating of soda bicarbonate on the filters; the same goes for pool lime on ultrafine dust or fume. In these applications, a substantial filter cake provides ultrafiltration. Using a modular setup with off-line cleaning is a good idea on these continuous bag coating applications.

For fabric filter collectors that use a precoat (such as for acid gas and/or mercury control), the air-to-cloth ratio for a reverse-air type collector is typically 2.5 to 1 or lower. If the collector is of pulse jet type, then the A/C ratio is about 3.5 to 1 or lower. The goal is often to keep the inlet flange to outlet flange pressure drop below 6–7 inches water column since the resistance of the collector impacts on the energy required to move the air through the system. In addition, with precoated type applications, the designer chooses the filter medium for efficient filter cake release. Various surface treatment techniques are used to tailor the filter cake retention to suit the application.

Air-to-cloth ratios are only guidelines. Many other factors affect performance. For example, the aspect ratio evaluates air-to-cloth ratio as it relates to dirty air velocities between filters in short or tall form collector. It is an especially important consideration because high velocity in low-inlet designs will not allow dust/particulate to drop down into the hopper.

7.6 Operating Suggestions

It should be obvious from the previous comments that to operate a fabric filter collector efficiently, it must first be sized correctly and then operated so that the collected dust (particulate) is removed properly. The mechanism to remove the particulate from the medium and the mechanism to remove the particulate from the hopper must be kept in good operating condition. If a shaker-type collector is used, the mechanical mechanism to shake the bags should be inspected and kept properly lubricated. If a reverse air-type unit is used, the reverse air isolation dampers and their actuators should be periodically inspected and maintained. These dampers and valves are critical to the reverse air's proper operation. If a pulse-type collector is used in cold climates, the compressed air supply should be conditioned or dried so that the fittings and valves do not freeze. The pulse timer (usually electronic) should be protected from voltage spikes so that its timing circuitry remains operable.

If the collector is used on a hot source containing acid gases (such as SO_2 and HCl) and periodically is shut down, the collector should be thoroughly insulated, and hopper heaters installed as needed. Some collectors utilize hot-air heating systems that recirculate air in the baghouse to uniformly distribute the heat. Failure to do so allows the baghouse environment to pass below the acid dewpoint, which causes localized corrosion and damage.

For pulse-type collectors, various Venturi and cage materials of construction (MOC) are available. These include coated Venturis, alloy wire cages, and so on. If the application is corrosive, attention should be paid to the MOC of the Venturis and cages. If the dust is explosive, special bags with grounding wires can be installed. Obviously, the grounding system should be inspected often to make certain that it is operating as intended.

For a hopper discharge problem in which the dust tends to bridge over the dust outlet, bin activators (shakers) or acoustic horns can often be used to break up such bridging. Usually, a continuous flow of dust out of the collector is better than an "accumulate and dump" type scenario.

On pulse type units, the pulse headers can often be removed from the top (clean air side), but space must be allowed for their removal. Some designs allow for the headers to be pulled out laterally. Again, one must plan for their removal.

If a bag breaks, you usually are in trouble. For that reason, various vendors offer broken bag detectors that scan the CAP for signs of particulate. If a broken bag is found, it is not uncommon to replace the row in which that bag was found as well as the adjacent rows. When one bag fails, it usually is a sign that others will follow.

To reduce bag injury upon installation, the bag tube sheet holes should be thoroughly deburred. New bags should be installed vertically (if that was their original orientation), not on an angle. This prevents the cage from chafing the medium.

On pulse-type units, the bag pulse frequency and duration should be carefully selected (most vendors have their required settings based upon experience). The pulse start sequence can often be initiated by a pressure switch so that a precoat of particulate can build up first. Every pulse in some small measure reduces the life of the bag, so pulsing should be done only as needed.

Shaker-type collectors often have media tensioning devices that require initial setup and checking. The collector manufacturer asks that these measures be followed to get the most use from the media. Unfortunately, these details are often overlooked.

On applications involving the temporary retention of a filter cake, over pulsing or over shaking the medium can shorten the life of the medium.

Fabric filter collectors provide excellent service when properly applied to the application and when they are operated as the designer intended.

8

Fiberbed Filters

Joe Mayo

Air-Clear, LLC, Frazer, Pennsylvania

8.1 Device Type

Fiberbed filters are specialized filtration devices that are primarily designed to coalesce and capture liquid contaminants such as acid mists and aerosols, the viscosity of which is low enough that they flow or can be made to flow from the fiberbed surface.

The design gets its name from the medium used. It consists of micron-size fibers that are compressed tightly in a mat or bed, which provides the surface area and gas path thickness needed to capture the pollutant.

These designs are somewhat related to filament/mesh scrubbers in which they utilize target fibers in a wet environment. The fiberbed filter fibers, however, are in 5–15-μm diameter range, or a fraction of the diameter of the filament- or mesh-type scrubbers. The fiber spacing is therefore closer in a fiberbed filter and, in general, it can remove smaller diameter aerosols.

Figure 8.1 shows a cutaway view of a fiberbed filter unit. The individual filters (sometimes called candles, given their shape) are mounted on a tube sheet in either a hanging or sitting position. The unit shown shows them hanging from a tube sheet. The small J-shaped pieces under each candle are liquid traps that allow the liquid to drain but prevent gases from bypassing the filter.

8.2 Typical Applications and Uses

The following are brief descriptions of common fiberbed filter applications. With one exception, they all involve the collection of liquid droplets. In general, if the exhaust stream is wet or the particles in the exhaust are liquids, or if a high-efficiency filter that can withstand a high-pressure drop is required, then fiberbeds are a potential control option (see Figure 8.1).

FIGURE 8.1
Cutaway of fiberbed filter (Air-Clear, LLC).

8.2.1 Acid Mist

Collecting acid mist was the first significant commercial use of fiberbed filters and is still the largest application for them. Most sulfuric acid manufacturing plants use fiberbed filters in the absorbing and drying towers to remove SO_3 and liquid acid mist from the air. Fiberbeds are also used to remove residual mist in the exhaust of wet scrubbers, particularly hydrochloric acid scrubbers, because the reaction with the scrubbing liquid can be violent and creates a visible emission from the scrubber. These are typically cool and clean applications, requiring no prefiltration or cooling.

If additional fiberbed surface area is required, a nesting or concentric-type filter can be built. In these designs, as shown in Figure 8.2, a fiberbed is mounted within another fiberbed, thus increasing the face area of the medium and slowing the gas velocity. The reduced gas velocity is said to improve the capture of aerosols and mists.

8.2.2 Asphalt Processing

Equipment used in asphalt processing include coaters, saturators, converters (blow stills), storage tanks, and truck loading and unloading facilities. The coaters and saturators used in roofing manufacture often have solids that

FIGURE 8.2
Filter within a filter (Monsanto Enviro-Chem Systems, Inc.).

must be prefiltered before the fiberbeds. Saturator exhaust may also require cooling. Tanks and loading racks usually achieve adequate cooling through radiant losses in the ductwork and have little solid particulate. Asphalt converters are also relatively free of solids but may require cooling. Such a unit is shown in Figure 8.3.

8.2.3 Plasticizer/Vinyl/PVC Processing

Vinyl and PVC processing, such as calendaring, coating, and curing operations, emit oily plasticizers and other materials that can cause a substantial exhaust stack plume. While oven exhaust must usually be cooled to condense the vapors, coater and calendar emissions are often captured by canopy hoods that draw in ambient air that cools the exhaust. Prefilters are usually not required.

FIGURE 8.3
10,000 acfm system with prefilters (Air-Clear, LLC).

8.2.4 Coating/Laminations

Many coating and laminating processes, especially on fabric and vinyl, create visible emissions that fiberbed filters can effectively control. The emissions are typically generated during the drying and curing phase of the operation, so the exhaust is hot and usually requires cooling to condense the vapors. The cooling coil housing is on the right side in Figure 8.4.

8.2.5 Electronics

Electronic component manufacturing, such as solder leveling, can create oil mist from the fluxes used. Fiberbeds can also be used as point source collection for acid mists, reducing the load on house scrubbers and reducing salt

FIGURE 8.4
With prefilters, water cooling coils on curing ovens (Air-Clear, LLC).

formation in the ductwork. Materials of construction must be carefully chosen because many of the materials are potentially corrosive.

8.2.6 Textile Processing

Textile tenter frame ovens and dryers can emit a mixture of pollutants
including oils, resins, waxes, tars, and various solids, producing a prodigious
stack plume. This hot, dirty exhaust requires both cooling and prefiltration.
The mineral oil-based emission from a tenter frame can be collected using a
fiberbed as shown in Figure 8.5. Note the induced draft fan and exhaust duct
located to the right of center.

8.2.7 Metal Working

Coolant and oil mists are often generated by the high temperatures at the tool
working surface. Grinding operations usually require prefilters to protect the
fiberbeds from swarf. Such a system is depicted in Figure 8.6. A water wash-
down system is sometimes used to flush the interior of the system free of the
water-based coolant to avoid long-term growth of bacteria inside the system.
In general, when insoluble particulate or fibers are present, a prefilter should
be used.

FIGURE 8.5
30,000 acfm system on tenter frame (Air-Clear, LLC).

FIGURE 8.6
1000 acfm on five-station machining center (Air-Clear, LLC).

8.2.8 Lube Oil Vents and Reservoirs

Oil lubricating systems, such as used on gas and steam turbines, often emit oil mist due to the hot oil returning from the turbine. No cooling or prefiltration is usually required. The compact cylindrical design of the fiberbed shown in Figure 8.7 makes these easy to install on lube oil vents. These also serve to recover oil and thereby reduce maintenance expenses. A similar configuration is used on ocean-going naval vessels for crankcase ventilation systems (mentioned in Section 8.2.10).

8.2.9 Incinerator Emissions

Incinerators that burn toxic, hazardous, or radioactive materials may produce submicron particles that must be controlled. Typically located downstream of a wet scrubber, the fiberbeds can be made of polyester or other materials that can be completely incinerated to dispose of spent filter media.

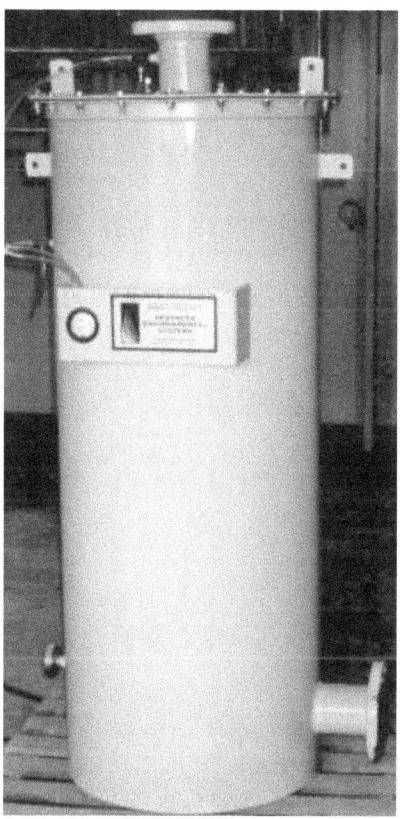

FIGURE 8.7
300 cubic feet per minute (cfm) oil vent unit (Air-Clear, LLC).

8.2.10 Internal Combustion Engine Crankcase Vents

Internal combustion engines have crankcase oil mist emissions due to blow-by around the piston rings that are economically controlled by fiberbeds. This application is like lube oil reservoir vents (see Figure 8.8).

8.2.11 Precious Metal Recovery

Process catalysts such as palladium gauze in nitric acid manufacturing can be lost into the process stream. The high temperature stability and structural strength of fiberbeds make them ideal for recovering these valuable metals. This is the unusual case of fiberbeds being used to collect solid particulate.

FIGURE 8.8
Packaged fiberbed with prefilter (Air-Clear, LLC).

8.2.12 Vacuum Pumps

Vacuum pumps mechanically generate oil mist during their operation and, unless they are evacuating furnaces, are usually cool. Some applications such as silicon crystal growing contain solid particulate (silicon dioxide) and thus require prefiltration. The prefilter in the unit in Figure 8.9 removes the particles that could plug the main filter.

Another method of prefiltering involves encasing the main fiberbed candle with a removable outer filter. The man in Figure 8.9 has these prefilters draped over his shoulder. Note the retaining cage to the left.

You would not be well advised to use fiberbed designs to clean gas streams containing inert particulate or liquid aerosols that do not flow by gravity or

FIGURE 8.9
Removable filter medium (Monsanto Enviro-Chem Systems, Inc.).

resist water or solvent washing. Solid particulate can blind the filter. This problem is often solved using prefilters or prescrubbers.

8.3 Operating Principles

A fiberbed filter uses a densely packed bed of microfibers placed in the path of the contaminant gas stream. The fibers become obstacles that the gas and contaminants must traverse. The closely spaced arrangement of the fibers improves the probability that a contaminant, such as a liquid aerosol or acid mist, will adhere to and coalesce upon the fibers. As this procedure progresses, the liquid builds up to a point at which it can drain by gravity.

8.4 Primary Mechanisms Used

Fiberbed filters operate using three basic mechanisms: impaction, interception, and Brownian diffusion. Impaction and interception are popular mechanisms used in various gas cleaning devices. Brownian diffusion, however, is primarily found in use in fiberbed collectors.

As air containing particulate flows through a filter, the air flows around any obstacle (such as a filter fiber) that is in its path. But a particle with sufficient mass and momentum (such as a 5-μm particle) will not. Instead, the particle's inertia will cause it to continue along its original path until it strikes a filter fiber and is collected. This is termed as *impaction*.

Somewhat smaller particles, those in the range of 1–3 μm, are collected by *interception*. Because these smaller particles have less mass and therefore less momentum, they tend to follow the airstreams around a filter's fibers. However, they can stray a bit from the normal streamline and can graze the side of a fiber and be collected.

Small particles (less than 1 μm) have little mass and as a result follow the air as it winds its way through a filter. These particles have substantial random motion, called *Brownian diffusion*, due to collisions with nearby air molecules. This almost vibratory motion allows them to move independently of the motion of the bulk airstream. Like gases and chemical solutions, the particles tend to migrate or diffuse from areas of high-particle concentration to areas of low concentration. As the particles contact the filter's fibers and are collected, the concentration in the air near the fiber surface goes to zero. This cycle of diffusion and collection drives the removal of the submicron particles.

Because slower operating velocities increase the time available for the diffusion to occur, fiberbeds have infinite turndown capability. As the collected particles coalesce into larger droplets on the fiber surface, they drain from the filter by gravity.

One of the pioneering fiberbed designs was the Brinks mist eliminator. Manufactured by Monsanto Enviro-Chem, the fiberbeds are made from glass or polymer microfibers often in the form of candles. Figure 8.10 shows a Brinks fiberbed mist eliminator.

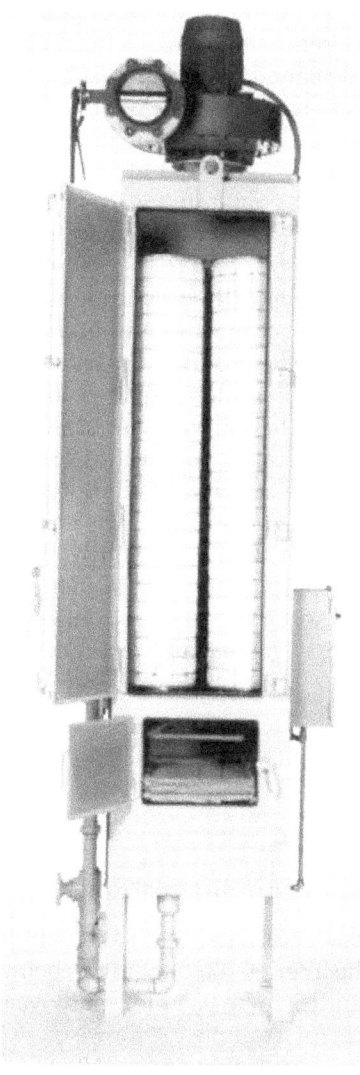

FIGURE 8.10
Brinks mist eliminator (Monsanto Enviro-Chem Systems, Inc.).

8.5 Design Basics

Fiberbed filters operate at inherently low vapor velocities both to maximize performance and to minimize pressure drop. Face velocities of 0.5 ft/s or less are common. In general, the higher the liquid loading, the slower the required gas velocity. This often results in a significant number of candles for even low gas volume applications.

An inner and outer cage usually supports each candle. The cage may be made from metallic or nonmetallic mesh of high open area. These cages retain the compressed fiber material that is captured between the cages. The outer cage is typically designed to be removed for repacking.

Because there is a time delay within which the captured aerosols or mists coalesce, a new candle can take several hours to wet out. The fiberbed achieves its best performance after the fibers are coated with a film of liquid (provided by either the contaminant itself, an irrigation system, or an administered fog or mist). It is not uncommon for a fiberbed to exhibit low efficiency when new.

The candles themselves typically use a mounting flange that is bolted to the tube sheet. The tube sheet must be designed for the laden weight (wet weight) of the fiberbed candle, not just its dry weight. Given that the tube sheet is weakened by the openings required for the candles, special care must be taken in stiffening the tube sheet sufficiently.

The accumulated liquid must be given a path through which it can drain, otherwise the candle retains the liquid, and its effective open area decreases. Small J-shaped traps are often used on each individual candle to allow the liquid to drain, while preventing liquid from bypassing the candle and reducing efficiency. These traps must be filled with liquid before operation. They must also be of sufficient depth to seal at the maximum anticipated pressure drop. This usually results in a seal leg of 12–18 inches overall length.

8.6 Operating/Application Suggestions

Fiberbed filters can provide reliable service on applications where the contaminants flow from the filter media rather than being retained on the media. It is not unusual for candles to be used for many years without replacement in acid recovery service, for example.

Following are some measures that can be taken to maximize the useful life of a fiberbed system.

8.6.1 Filter Cleaning

Fiberbed filters cannot be cleaned in the traditional sense, as their structure is delicate and easily damaged. Accumulations of soluble materials such as salts can be removed by irrigating or flushing the filter with water or another suitable liquid. Waxes and tars can often be removed by heating the filters indirectly through injection of low-pressure steam into the filter vessel. Several hours of heating (with the system shut down) can liquefy waxes and other materials, enabling them to drain from the filters. Detergent sprays can sometimes also be used to flush insoluble materials from the filters, but this procedure usually must be done daily to remove the insoluble materials before they accumulate.

8.6.2 Fiberbed Filter Life

Fiberbed life in any given application is determined by four major factors. These are the concentration of foulants (materials not draining from the filters), fiberbed surface area, starting pressure drop of the filters, and the pressure available from the exhaust blower. As foulants build up on the filters, the pressure drop across the filters increases. When the limit of the fan static pressure capacity is reached, the filters must be replaced.

While the foulant concentration cannot be changed, the other three items can. Increasing the number of filters both increases the surface area and decreases the pressure drop. Increasing the pressure capability of the fan further increases fiberbed's life, because this allows the pressure drop to increase further before reaching the fan's limit.

Because all the pressure capability of the fan is not needed when the filters are clean, a damper or variable frequency drive (VFD) is used to control exhaust flow. A damper would be mostly closed at startup, and a VFD would be running the fan at a low rpm. As the pressure drop increases, the damper is opened or the VFD speeds the fan up to maintain flow. When the damper is fully open or the fan is running at maximum speed, the limit of the system has been reached and the filters should be replaced.

With all these variables it is difficult to generalize, but in fiberbed systems professionally designed for the application, filter life is usually anywhere from 2 to 6 years.

8.6.3 Fire Protection If the Contaminant Is Combustible

Fiberbeds are often used to collect combustible contaminants. This can be accomplished safely if a few precautions are taken.

Fire protection is an important part of any system collecting combustible materials. Fires usually begin upstream of the fiberbed system, for example, in a direct-fired oven. If the fire spreads to the oil-saturated fiberbed filters, they may catch fire. Burning fiberbeds is difficult to extinguish because their thick walls act as an insulator.

Water sprinklers are the best choice for fire protection, because they can be used to flood the fiberbeds. Water not only extinguishes the fire but also carries away heat, reducing the possibility of reignition. Isolating the fiberbed chamber and smothering the fire with steam or carbon dioxide can also be used. In any case, the filters should be removed from the vessel as soon as possible after a fire and monitored to ensure they do not reignite.

Fire detectors are quite useful in minimizing fire damage. They should be located on the inlet and the outlet to the system and should be tied into the control system to shut down the system fan (to reduce the available oxygen), sound an alarm, and activate diversion dampers if used. They are available in a variety of temperature ranges and should be selected based on the maximum temperatures expected in the application to avoid unnecessary shutdowns.

Fire dampers can also be used to minimize the spread of a fire. The damper is located on the inlet to the fiberbed system and closes when temperatures indicative of a fire is detected. This stops the flow of air through the filter vessel, which can occur even if the exhaust fan is shut down, due to chimney effect.

9

Filament (Mesh Pad) Scrubbers

9.1 Device Type

Filament or mesh pad-type absorbers have proven themselves highly effective in the absorption of water-soluble gases. When constructed in a sufficiently dense medium panel, they can be used for the collection of airborne bacteria and spores. These devices typically use woven or layered filamentaceous mesh layers onto which a spray of liquid is administered. The contaminant gases, particulate, or both pass through these layers of mesh wherein the contaminant and liquid are brought into intimate contact, thereby promoting gas absorption. Various vendors have developed proprietary designs of this generic type with hundreds of successful installations.

9.2 Typical Applications

Filament and mesh pad-type scrubbers are often used to collect inorganic acid vapor emissions from process reactors or storage tank vents. They are also used after particulate removal devices to enhance the absorption of gases. Laboratory hoods and point of use scrubbers often use wetted filament or mesh pad scrubbers.

They are often used on large gas cleaning systems, for example, for acid concentration and capture. Figure 9.1 shows a multistage wet scrubber on a superphosphate fertilizer plant for the recovery of fluosilicic acid. This system consists of a Venturi scrubber for particulate control, a multistage-wetted Kimre pad unit for stepwise acid gas concentration, and a preformed spray scrubber for polishing the emission using pond water. The water flows upstream from the preformed spray scrubber to the filament-type absorber. The fluosilicic acid is concentrated in three-sprayed stages in the filament-type scrubber, which is housed in the rectangular box on the left (ahead of the fan). The mesh pad-type scrubber is shown in the foreground of Figure 9.2.

FIGURE 9.1
Multiple-stage superphosphate plant system (Bionomic Industries, Inc.).

Another common application is in the control of gases and mist from chemical reaction vessels. These vessels may incorporate an acid (or base) feed system that injects the chemical(s) into a vented, agitated tank. Given the agitation and chemical reaction, some absorbable gases may be evolved along with droplets. Often, the mesh pad-type device drains back into the reaction vessel, thus serving to capture and recycle the mist and chemicals.

Filament or mesh pad-type scrubbers are generally not used where insoluble particulate is present or where a precipitate can form during chemisorption.

FIGURE 9.2
Crossflow mesh pad-type scrubber in foreground (Bionomic Industries, Inc.).

9.3 Operating Principles

In these devices, the mesh serves several purposes. It extends the liquid surface in a compact space, thereby providing the liquid surface area required for effective gas absorption. It also helps hold up the liquid to allow sufficient time for the contaminant gas to diffuse to the liquid surface and be absorbed. The compact nature of the mesh also reduces the path length (distance the gas molecule must travel to the liquid surface), thereby, in theory, enhancing the rate of diffusion per unit volume of the media. Because the media are typically layered, the gas molecules are caused to move back and forth through the media, thereby increasing the probability of absorption into the liquid surface.

9.4 Primary Mechanisms Used

The filament or mesh pad-type mass transfer device is primarily used for gas absorption where the particulate loading is low. The mechanisms used are diffusion, gas absorption, chemisorption (if the liquid contains a reactive chemical), condensation (if the liquid is colder than the saturation temperature of the gas stream), interception, Brownian motion, diffusiophoretic and thermophoretic forces, as well as impaction if larger particulate is present.

Given the narrow openings between filaments or mesh layers, these devices are typically not used where solid particulate is present. The liquid spray or accumulation of liquid on the mesh tends to draw particulate into the mesh where the particulate can become lodged and difficult to remove. They, however, offer good removal characteristics for acid aerosols and other flowable liquid particulate down to about 1–2 µm aerodynamic diameter.

9.5 Design Basics

Filament or mesh-type devices can be configured for horizontal counterflow gas/liquid interception, or crossflow wherein the gas and liquid move concurrently at least for a portion of their movement.

The liquid is sprayed at a rate of 0.5–4 gpm per square foot of media surface. The design face velocity of the medium is dictated by the allowable pressure loss. Gas speeds of 2–6 ft/s are commonly used. The pressure loss per medium stage can vary from less than 1-inch water column (w.c.) for a loosely woven pad to over 6 inches w.c. for a multilayer, compressed pad.

The simplest filament or mesh pad -type collector is a wetted mesh pad. Figure 9.3 shows a mesh pad removed from a vessel. This pad can be used in a vertical gas flow or crossflow arrangement. If the gas moves vertically, the pad can be sprayed from the underside to flush away water-soluble particulate or help drain away dissolved contaminants.

Figure 9.4 depicts a crossflow multistage filament-type gas absorber supplied by Kimre, Inc. for the absorption of soluble gases, in this case inorganic acids. The typical gas inlet velocity is 40–55 ft/s and the vessel velocity is 2–6 ft/s given the vapor loading and amount of sprayed liquid.

Figure 9.5 shows another unit of this type wherein multiple stages of proprietary mesh pad layers are used. Basically, the more open multilayer mesh is used first followed by increasingly more dense mesh units. Two or three stages of carefully selected mesh types are commonly used.

Another interesting and effective type of filament-type collector, made by Misonix, Inc., is shown in Figure 9.6. This one uses a sandwich of square cloth mesh layers (much like coated window screen) that is pressed together in a proprietary fashion to create a high-density media panel. The proximity of the fiber strands in the panel greatly shortens the diffusion path and high mass transfer per unit volume results. This unit is equipped with a fog spray system (to the left) mounted ahead of the collecting pads (in the chamber to the right). The fog increases mass transfer and preconditions the gas stream before the media pads.

FIGURE 9.3
Multilayer mesh pad (Kimre, Inc.).

FIGURE 9.4
Crossflow scrubber (Kimre, Inc.).

FIGURE 9.5
Multilayered absorber module (Kimre, Inc.).

These types of collectors have also been used to control dopant gases used in the manufacture of semiconductor materials. These gas flows are typically exceptionally low (a few liters per hour), however the gases can be difficult to scrub. Figure 9.7 shows a packaged unit that includes a recirculation pump and chemical neutralization system.

Sprayed with a suitable biocide, these devices can also trap and control airborne bacteria and spores. They are therefore often used for laboratory hood applications.

Crossflow filament-type units sometimes have the medium inclined on a 10- to 15-degree angle with respect to the gas flow. This is done because

FIGURE 9.6
Crossflow scrubber with Sonimist chamber (Misonix, Inc.).

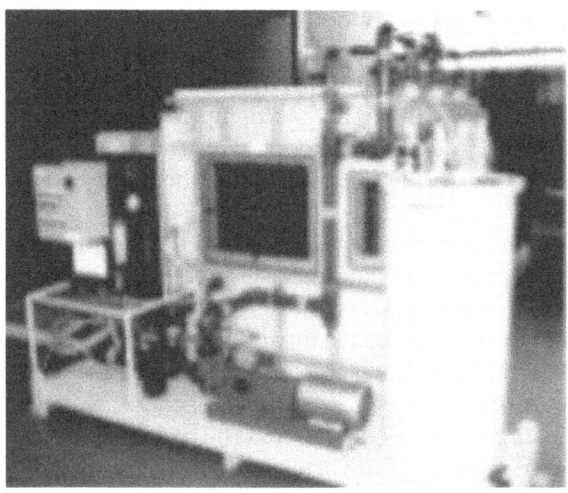

FIGURE 9.7
Packaged scrubber for semiconductor application (Misonix, Inc.).

the sprayed liquid does not take a purely vertical (downward) path. The gas velocity pressure tends to push the liquid in the direction of the gas flow. To help keep the gas in contact with the liquid, the medium is thus inclined so that the gas helps hold the liquid in the medium.

Crossflow units are often sprayed both from the front and from above. Interstage pads on multistage units are often run passively, that is, no sprays, but the upstream pad is run at a higher speed, so some of its liquid agglomerates and entrains from the lee side of the pad, thereby wetting the downstream pad without overloading it with liquid. The first pad, in other words, serves as an agglomerator for the downstream pad.

9.6 Operating Suggestions

Because the various mesh types are proprietary, it is suggested that you contact the specific vendor regarding application and operating suggestions.

The filament-type media are easily made into removable panels or pads, therefore service pull space should be designed into any installation. It is not uncommon to have a spare pad assembly ready and waiting for transfer if the primary pad plugs. The superphosphate reference above, for example, has a built-in access rail mounted above hinged service doors in the roof of the scrubber housing. The pads were designed to hang, much like pants folded over a hanger. The entire pad assembly can be pulled vertically upward out of the scrubber and moved away from the device while a new pad is installed.

Vertical flow mesh pad devices can sometimes be cleaned in place if sprayed from the bottom at the rate of 1–2 gpm/ft^2 of frontal surface area. They rarely can be cleaned on the run by back spraying from above because the pad acts as a check valve. The gas rising through the pad prevents the liquid from draining, and the pad floods. This causes entrainment. The pads can, however, often be cleaned in place when the gas flow is zero, that is, the scrubber is off-line.

Because the pads typically get heavier after use (through particulate or scale buildup), the panel size should be selected based on the laden weight of the pad, not its dimensions. Access doors should obviously be sized to allow the largest panel to be removed easily and safely. Hold-down bars should be used to secure vertical gas flow-type designs because the pads can tend to lift since the dry pad's open area decreases as it becomes wetted.

A simple pressure drop indicator across the pad can provide a good indication of the pad's condition. The various vendors have accurate dry and wet pressure drops for given gas flow characteristics so pressure drop increases can be used as an indicator of residual open area of the media. Quite often, this pressure drop is used to trigger a cleaning spray.

On acid gas applications, the vendors often suggest prewetting the pad(s). The vendor's recommendations should be carefully followed to achieve the best performance.

10

Fluidized Bed Scrubbers

10.1 Device Type

Usually, the term *fluidized* has been applied to the two-phase mixture of gas and solids. Fluidized bed boilers use, for example, an agitated mixture of fuel (coal), combustion gases, and sometimes lime or limestone to enhance combustion while reducing emissions. The gas is injected into a mobile bed of solids. A two-phase mixture of liquid and gas can equally be called fluidized. This technique injects the gas into a mobile, agitated, zone of liquid. Because the speed of dissolution of gases into liquid is typically enhanced by stirring, the agitation in these designs is intended to increase the speed of mass transfer. The vessel velocities are therefore typically higher than other types of absorbers.

Fluidized bed scrubbers can be divided into three major categories:

1. Mobile media-type units
2. Ebulating bed-type designs
3. Swirling, Coriolis induced, or co-mixing type.

10.2 Typical Applications and Uses

Fluidized bed-type scrubbers are used primarily as gas absorbers where particulate is also present that could plug other absorber designs (such as packed towers). The particulate may arrive in the gas or liquid stream or be a product of the reaction of the absorbed gas and the liquid.

They are noted for their compact size, low-to-moderate cost, and the ability to absorb gases while resisting plugging.

Common applications include the following:

1. Pulp mill bleach plant chlorine and chlorine dioxide control
2. SO_2 control using sodium hydroxide or sodium carbonate
3. SO_2 control using a slurry (lime/limestone, MgO, etc.)

4. Odor control (mercaptans, H₂S, etc.)

5. Gas cooling and condensing

6. Prescrubbing (ahead of other devices such as wet electrostatic precipitators)

7. Fluorine abatement (scrubbing with pond water in the fertilizer industry)

8. Stripping volatiles from dirty water

9. Acid gas control

10. Ammonia absorption

11. Bioslurry scrubbing

12. Humidifying biofilters

13. Where space is a premium

10.3 Operating Principles

The fluidized bed scrubber is related in many ways to the tray scrubber in that the fluidized bed scrubber is essentially a tray scrubber designed to operate at exceptionally high gas speeds. It is suggested that the reader also consult the tray scrubber chapter.

Mobile media type scrubbers include the universal oil products turbulent contact absorber (TCA) or ping-pong ball type design wherein a movable medium is supported above a support grid and below a bed-limiting grid. The TCA scrubber uses round balls that are agitated by the upward motion of the gases as the gases move through the vessel. The motion is intended to increase gas/liquid mixing and help keep the medium clean.

A more recent design is the Euro-matic UK Ltd Turbofill™ scrubber that uses ellipsoidal (egg)-shaped media. The skewed center of gravity of the medium causes the medium to exhibit nutation or oscillation as the gases pass through the zone. If solids are present, three phases can exist simultaneously (gas, liquid, and solid). These scrubbers also use support and bed-limiting grids whose openings are smaller than the medium size. The churning agitation of the medium helps keep it clean. Figure 10.1 diagrammatically shows the action of the medium.

Ebulating bed-type scrubbers are like the mobile media type, however they do not use media. *Ebulating* refers to the boiling-type appearance of the fluidized bed. The liquid bubbles randomly, much like a pot of boiling water. These designs incorporate perforated plates or mesh screen trays to provide high-velocity gas injection points that are used to fluidize the liquid and create the desired highly turbulent gas/liquid contact zone.

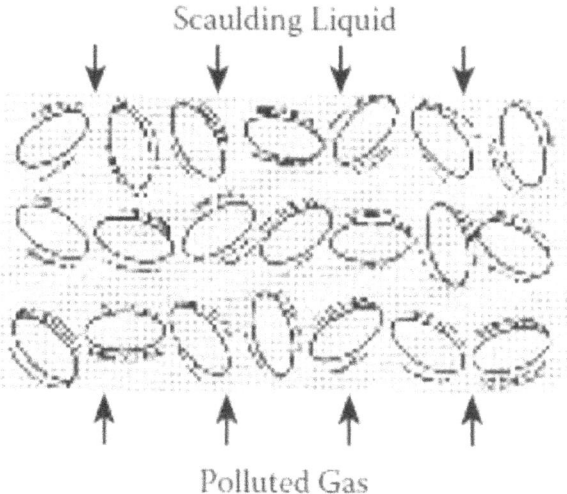

FIGURE 10.1
Nutating fluidized media (Euro-matic Ltd.).

The *sieve tray scrubber* is one of the oldest ebulating-type designs. The design consists of a vertical vessel in which at least one flat tray is installed perpendicular to the gas stream flow direction (upward). *Impingement tray* varieties, also of the tray scrubber family, use holes or perforations in the tray that are sufficiently small to allow the gas to pass upward through the hole but prevent the liquid from draining through. Figure 10.2 shows this type of scrubber. If the holes are enlarged, that is, the open area of the tray is increased, the gas velocity is insufficient to prevent the liquid from draining through the holes. The gas and liquid in effect compete for the same opening. These designs are often called *weeping sieve tray scrubbers*. The liquid drains through the holes, thus the term *weeping*. These are sometimes called counter-flow or dual flow trays because the gas and liquid pass counterflow through the same opening. Multiple trays can be used in one vessel, thereby repeating the fluidized zones until the proper number of transfer units (NTUs) is achieved. The liquid is generally introduced free flow either through low-pressure headers (with or without spray nozzles) or a weir arrangement like an impingement tray scrubber.

As attempts are made to increase the gas velocity through weeping sieve trays, the point is reached where the gas can pass upward through the opening, but the liquid cannot drain. This is called *flooding*, and the speed at which it occurs is called the *flooding velocity*. Experience has shown that just before flooding, mass transfer is at its greatest. Gas velocities more than flooding, however, usually cause a drop in performance and adverse conditions such as surging and dumping. *Dumping* is a condition where the

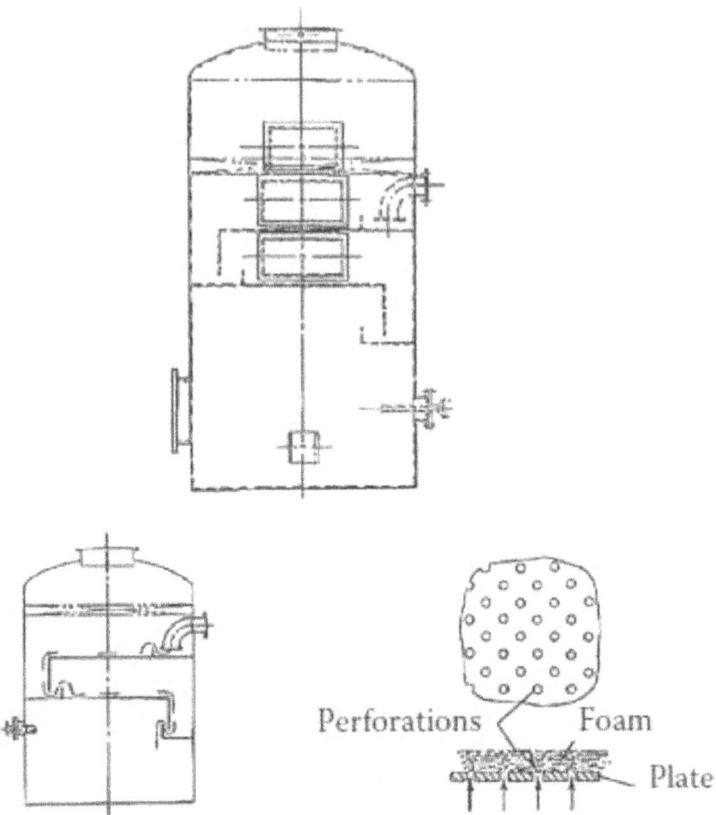

FIGURE 10.2
Impingement tray scrubber.

inventory of liquid above the tray or grid periodically dumps out of the ebulating zone.

Weeping sieve tray scrubber designers over time increased the tray hole size to allow smoother operation near flooding, but a point was reached where the hole opening was excessively large and did not afford adequate liquid coverage. Blowholes could occur where uncontacted gases could pass up through the tray/grid opening, thereby reducing efficiency.

In 1984, a U.S. patent was issued to an ebulating bed scrubber design that uses a grid mesh so open that it lacked structural strength and sagged (curved grid scrubber). The curved grid was shaped like the catenary shape of a hanging telephone. Figure 10.3 shows the patent drawing from this invention.

The design basis addressed the fact that as gas rises axially and vertically up a tower, it forms a velocity pressure profile. The velocity pressure

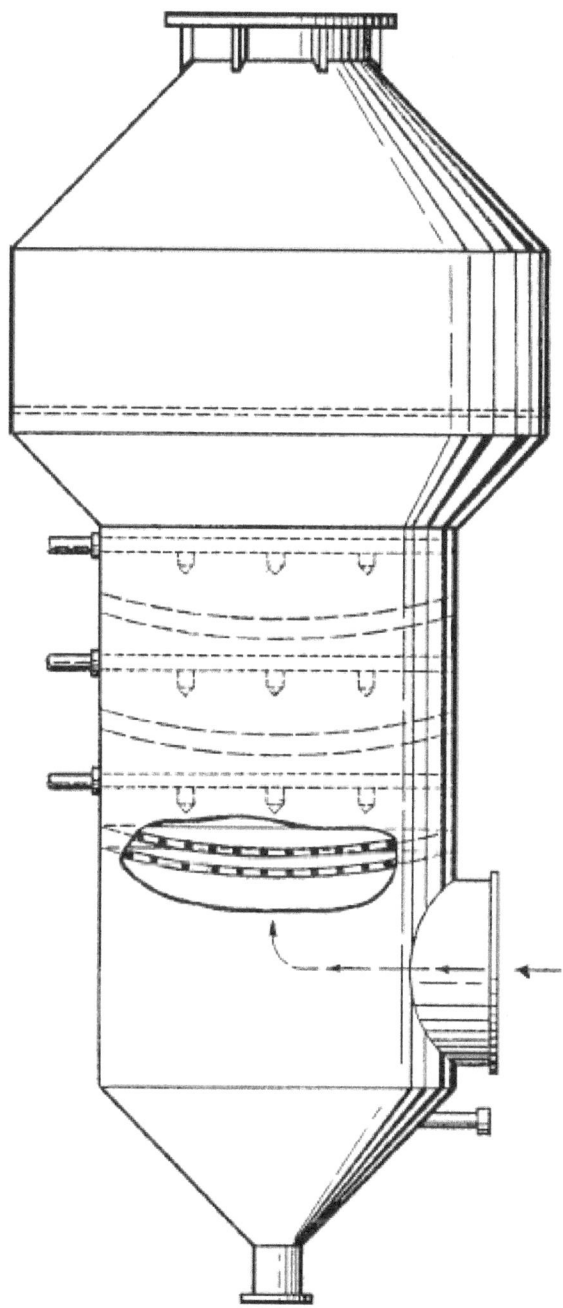

FIGURE 10.3
Ebulating bed scrubber (U.S. Patent Office, U.S. Patent 4,432,914).

profile represents the kinetic energy at any point on the curve (as opposed to the gas volumetric flow rate). The curvature of the grid used is the mirror image of this velocity pressure profile. It allows, therefore, a greater depth of liquid to form where the velocity pressure is the greatest, thereby making the application of kinetic energy more uniform and efficient across the diameter of the vessel. When liquid is dispersed above this grid, an ebulating zone is created. The zone increases in depth until the given gas velocity cannot support it any further. The liquid then starts to drain through the grid. The grid is uniform along its surface, and the gas has no preferred or directed path. These designs were sold under license by ChemPro (Fairfield, New Jersey), the Otto H. York Company (Parsippany, New Jersey), and others.

It became evident that on some of these installations the random bubbling of the bed was *too* random for optimum operation. Much like an overheated pot of boiling water, jets could erupt unexpectedly and spill over, causing upsets and reduced efficiency.

In 1999, a U.S. patent was issued on a fluidized bed scrubber device that both eliminated the media of the mobile media type and created a stabilized swirling rather than ebulating bed. This device is called the *RotaBed® scrubber*. It is marketed by Bionomic Industries. This design harnesses the Coriolis effect to create a stabilized, slowly rotating fluidized ebulating bed. A special swirl inducer and vortex finder as seen in Figure 10.4 were developed to create this desired action. The swirl inducer also provides structural support for the grid, a function totally lacking in curved grid designs. The special vortex finder was developed to provide a swirl pivot point about which the draining bed could pivot.

The gyroscopic stabilizing effect is much like that of a spinning top. Give a top a spin, and its angular momentum helps stabilize its rotation. Give the fluidized bed a spin and make it pivot about an axis, and greater stability is achieved. This free energy caused by the rotation of the earth helps impart a slow spin to the fluidized bed essentially without additional energy input. The swirl vanes are designed to get the rotation going and to help make the liquid draining more uniform. A slow swirling is desired, not a rapid one; otherwise, the liquid would be thrown to the vessel wall and become ineffective for mass transfer by reducing its surface area.

Rather than using the axial velocity pressure profile, the *RotaBed* scrubber creates a corkscrew-type gas pattern through the fluidized bed. This helical pattern increases the path length of the gas with the liquid, thereby improving mass transfer. Figure 10.5 shows a typical *RotaBed* grid complete with vortex finder (center) and the small vanes that impart the rotation. North of the equator, the vane pitch creates a counterclockwise rotation, and south of the equator a clockwise rotation is imparted.

Fluidized bed scrubbers often use quite simple liquid injection headers such as those shown in Figure 10.6. The view is looking down toward the grid, with the vessel on its side. Note that the headers are flanged bayonet

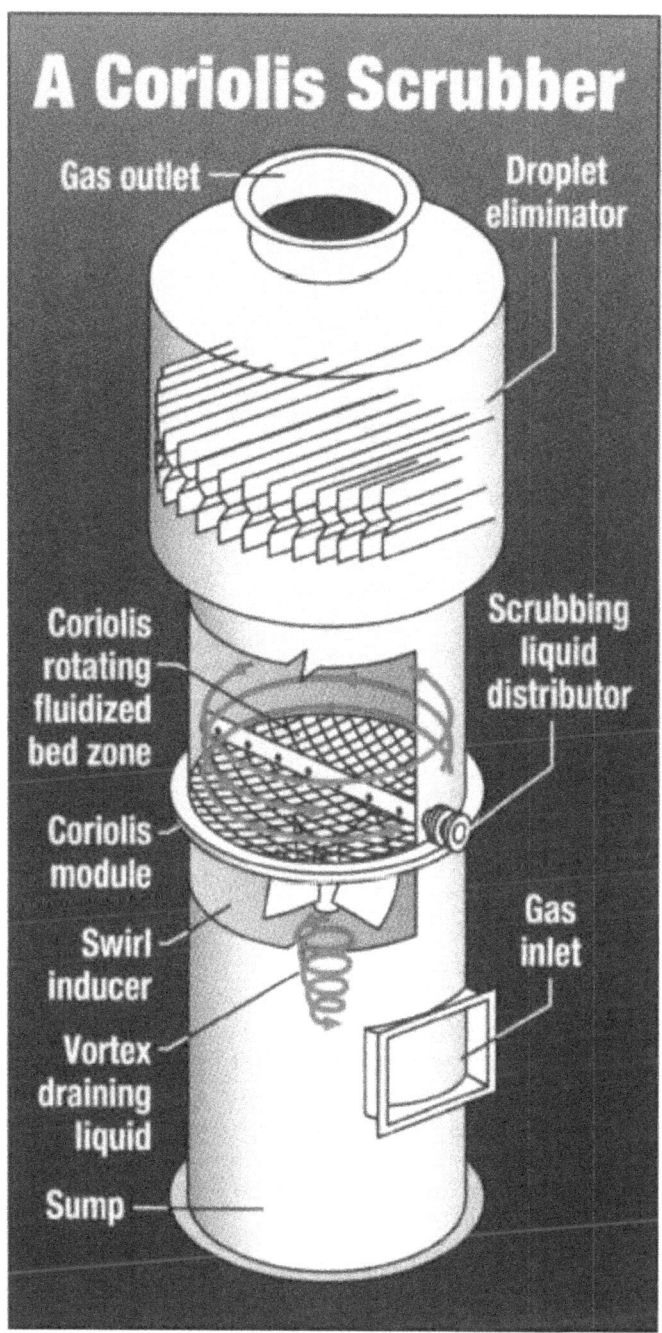

FIGURE 10.4
RotaBed scrubber (Bionomic Industries, Inc.).

FIGURE 10.5
RotaBed grid (Bionomic Industries, Inc.).

type, which can be retracted from the vessel. These headers are submerged in the scrubbing liquid during operation so the turbulent action of the liquid surrounding the headers helps keep them clean. The headers typically use low-velocity horizontal holes for liquid injection; therefore, their back pressure, that is, pumping pressure, is inherently low (less than 5 pounds per square inch [psig]). No spray nozzles (that could plug) are used.

10.4 Primary Mechanisms Used

These designs use the gas absorption principles described in Chapter 1 and elsewhere.

FIGURE 10.6
Header arrangement (Bionomic Industries, Inc.).

The typical NTUs available varies from 0.5 NTU/stage to approximately 2 NTUs/stage, depending on the application. The lower the solubility of the pollutant gas, the lower the NTU available. These designs rely on the rapid absorption of the gas followed by a rapid reaction of the gas with chemicals in the liquid.

For particulate capture, the primary capture mechanism is impaction. Particulate above 10 μm aerodynamic diameter can be removed at 80%–95% efficiency. There is a rapid drop-off in particulate removal efficiency below 5 μm diameter because these scrubbers are intentionally operated at low pressure drop (usually under 6 inches water column [w.c.]). As a gas cooler, the highly agitated bed creates shorter gas to droplet path lengths and affords superior application of diffusion and phoretic forces.

10.5 Design Basics

Typical gas inlet velocities are 45–55 ft/s. Vessel velocities vary from approximately 8 to 10 ft/s for the mobile bed scrubber design to 18–30 ft/s for the RotaBed scrubber design. At the droplet removal stage, the gas velocity is reduced to 10–12 ft/s to accommodate a chevron or spin vane droplet eliminator. If a mesh pad is used, the gas velocity is decreased to 8–10 ft/s. Gas outlet velocities are the same as the inlet if the cleaned gases proceed to downstream equipment (such as a fan). If a stack is mounted on top of the scrubber, gas velocities of 35–40 ft/s are common.

Liquid header speeds are usually 2–6 ft/s velocity with header pressures of under 3–5 psig. Some designs permit the use of liquid headers at each stage, thereby allowing the adjustment of the scrubber chemistry within the scrubber at each stage.

Pressure drops range from approximately 0.5 inches w.c. per tray/grid to more than 6 inches w.c. per grid, depending on the fluidized bed depth. Typical pressure drops are 1–2 inches w.c. per grid or tray stage, with mobile bed scrubbers demonstrating slightly higher pressure drops.

Mobile bed scrubbers can use mesh pads, chevrons, or packed sections for droplet control. In the curved grid scrubber, which operates at a higher vessel velocity, droplet control is most often by chevron; therefore, these vessels are usually greater in diameter at the top (chevron requires a lower face velocity) than where the grids are located. The RotaBed scrubber has been installed with chevrons in the expanded diameter upper stage or in a cross-flow chevron-type droplet eliminator mounted after the scrubber.

Maintenance is typically incredibly low for these designs. The attrition rate on the mobile media varies by application. The materials of construction of the mobile media are usually limited to thermoplastics. The designs devoid of internal media are suggested where overheating or erosive-type particulate is present.

Vessels may be made of any suitable formable material. Grid/trays can be made in any material that can be perforated or drawn into structurally sound wire.

10.6 Operating Suggestions

Fluidized bed scrubbers are best used where plugging resistance is of paramount importance in a gas absorption application. Although less effective for particulate control, they can be used to remove large (10 µm) particulate and where the inlet loading of particulate is less than 5–10 grains per dry standard cubic foot. If the particulate is difficult to wet (example: certain clays and powders), it is best to prescrub the gas using a device specifically designed for particulate control such as a Venturi scrubber.

The liquid headers in fluidized bed scrubbers often are low-pressure designs with port (hole) openings rather than spray nozzles. These header holes are typically more than ½ inch in diameter and represent the smallest opening through which any solid must pass. If solids are expected to agglomerate in the liquid circuit, these header openings can often be enlarged.

With most designs, it is imperative that these headers eject the scrubbing liquid horizontally rather than vertically. Fluidized bed scrubbers are essentially energy balances between the gas kinetic energy and that of the liquid. If one sprays the liquid downward, excessive energy can be imparted to the liquid, making it more difficult to fluidize. If the liquid is injected horizontally, the weight of the liquid is its principal vertical energy component and fluidization is much easier to accomplish.

Because this energy balance exists, the liquid flow must be initiated before the gas flow. Fluidized bed scrubbers are noted for their extremely low dry, that is, no liquid, pressure drop. Take away the liquid and one takes away much of the flow resistance. If a fan is used, loss of liquid can cause the fan to be unloaded and run out on its fan curve, producing excessive gas flow. This gas flow can sometimes overwhelm the droplet eliminator, leading to entrainment.

Many fluidized bed-type scrubbers therefore have interlocks in the control circuit that require the addition of the liquid first and then permit the fan to start. If the liquid is lost during operation (given a pump failure, etc.), the fan is momentarily stopped, and the pump is restarted. This allows the gas velocity to fall below the fluidization speed and keeps the gas flow within the range of the droplet eliminator.

Still others use a gas reflux system that pulls gas back from the stack to the scrubber or fan inlet (if the fan is located ahead of the scrubber). A modulating opposed blade damper in this line modulates based on scrubber pressure drop or source draft, automatically keeping the scrubber within design gas flow range. Systems of these types can control draft-sensitive sources to within a few hundredths of an inch of water draft.

The inherent mixing action in a fluidized bed scrubber can simplify chemical addition. Although chemical is often added in the recirculation pump inlet for mixing, these types of scrubbers often use direct injection of chemical into the scrubber headers or even into the fluidized zone itself.

Usually, about one-third of the recirculated scrubbing liquid is held up in this scrubber's contact zone. When the scrubber shuts down and airflow ceases, this liquid will fall; therefore, the scrubber sumps must be designed for adequate freeboard if overflow upon shutdown cannot be tolerated.

11

Mechanically Aided Scrubbers

11.1 Device Type

Mechanically aided scrubbers are defined as devices that use a moving mechanism in the gas stream to achieve the desired particulate or contaminant gas removal.

Mechanically aided scrubbers are generally used for dust control involving particulate larger than approximately 10 μm diameter and at low loadings (below 5 grains per dry standard cubic foot). They have been used to control dusts from loading and unloading facilities, fugitive dusts from storage facilities, wet coating operations, and numerous other applications. They were particularly popular in the 1970s in the mining industry for dust control where a wet product was required and are still used for that purpose. They are noted for their low cost, compact size, and reliability. Ever-tightening codes have shifted the focus to low- to medium-energy Venturi scrubbers on many applications that had been dominated by mechanically aided designs.

11.2 Typical Applications and Uses

These designs are used to control high-dust loadings of relatively large dust where it is desirable to recover the dust wet. Controlling dust from conveyors, blenders, mullers, mills, and other high-dust-loading sources are areas where this type of scrubber has been used. In these applications, the dust is usually returned to the process as wet slurry or is separated in a pond or clarifier.

They are effective on particles 10 μm or larger. They are not used as much where the particulate is hygroscopic and may buildup on the scrubber inlet. Similarly, they are not used for gas absorption, although they may be used ahead of a gas absorption device (such as a packed tower).

These scrubbers have been applied to the control of emissions from dissolving tank (smelt tank) vents in the paper industry. Some have been enhanced through the addition of a Venturi reverse spray scrubber ahead of the mechanically aided scrubber. Some recovery boiler suppliers routinely supply mechanically aided scrubbers for use on the dissolving tank vents. In recent years, packed towers have been added after the mechanically aided scrubber for additional gas absorption and odor control.

Their compact size and low cost make them attractive, where codes allow, for general dust-control applications.

11.3 Operating Principles

As was mentioned previously, it is believed that a given amount of energy input into a gas cleaning device is required to achieve a given amount of pollutant removal (the equivalent energy theory). This energy may be introduced using the gas velocity (produced by a fan or other prime mover), through a pump or other means or pressurizing liquid, or by a moving mechanical device.

The most common moving mechanism used in mechanically aided scrubbers is a rotating fan wheel or modified fan wheel. The wheel is usually sprayed with scrubbing solution and the liquid is shattered into the desired droplets. Locally, high relative velocities exist between the gas and the liquid so that impaction is enhanced. These designs basically include the fan with the scrubber, so no additional fan is needed.

Two popular mechanically aided scrubbers are the American Air Filter (AAF) W RotoClone series and the Ducon UW-4 arrangement. These use sprayed wheel-type contacting stages but approach the problem differently.

Figure 11.1 shows the AAF RotoClone unit. The Roto comes from the specially designed rotating element, and the Clone comes from the cyclone-type separation that was used. In this design, the gas stream usually is ingested into the rotating wheel where high local velocities and centrifugal force were applied, along with scrubbing liquid, to impact the particulate into the droplets. The scroll-shaped housing served to separate droplets from the gas stream. The type N RotoClone is used for higher efficiency dust control.

Please note the drag chain conveyor (extending off to the left) that is an integral part of the sump. This drag chain allows the continuous or periodic removal of settleable solids from the scrubber. This makes for a very compact arrangement.

Another variant is the type W RotoClone. Shown in Figure 11.2, the type W provides for direct injection into the specially designed fan wheel. In this model, the motor (seen at the right) is providing power to the wheel via a V-belt drive.

The Ducon UW-4 type scrubber uses a primary cyclonic separation stage followed by a wetted fan, followed by another cyclonic separation stage. The primary cyclonic zone is used to centrifugally separate large particulate and droplets and thereby reduce the loading to the wetted fan. The wetted fan

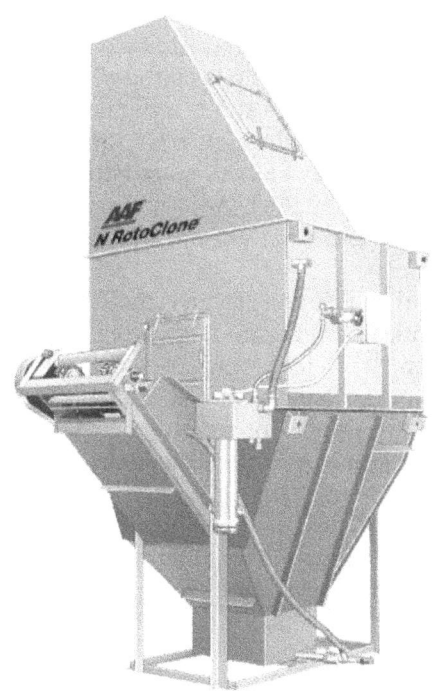

FIGURE 11.1
AAF RotoClone type N (American Air Filter).

FIGURE 11.2
Type W RotoClone (American Air Filter).

was sometimes preceded by a spray duct to add to the droplet loading and liquid/gas surface area. The wetted fan could be equipped with sprays both at the fan eye and in the housing to help keep the wheel clean. The spray regimen leaves the wetted fan discharge and is separated in a cyclonic separator. The captured liquid is returned through a trap to the primary separation stage or is diverted out of the vessel to a separate drain.

There are modified versions of the UW-4 type scrubber supplied by a variety of vendors. There were literally thousands of these scrubbers sold; some are running to this day.

An interesting mechanically aided design is the T-Thermal Hydrop® scrubber. Shown diagrammatically in Figure 11.3, the gas stream enters through a wetted approach section (at the top) and proceeds to a special inlet duct that discharges into the rotating wheel section of the device, where the stream is subjected to centrifugal and shear forces, causing the particulate to be combined with the injected liquid. A cyclonic separator is used to separate the droplets from the liquid stream.

Other mechanically aided designs have been the subject of experimentation including those using sonic pulses, microwaves, vibrating elements, or combinations thereof. In each case, some external force other than simply fan velocity or pump pressure supplements the total energy input.

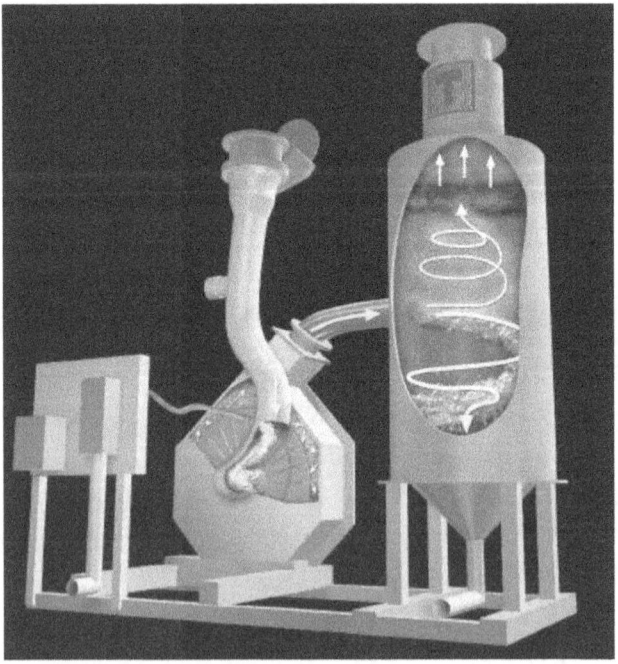

FIGURE 11.3
Hydrop scrubber (T-Thermal Company).

FIGURE 11.4
Rotary scrubber (TREMA Verfahrenstechnik GmbH).

The mechanically aided rotary scrubber in Figure 11.4 supplied by TREMA uses tangential gas inlets to provide prescrubbing. They can be seen near the base of the unit.

11.4 Primary Mechanisms Used

The primary particulate capture mechanism in these devices is thought to be impaction, given the high relative velocities between the gas and liquid. Interception could also play an important role in mechanically aided scrubbers.

Because the agitation is sustained only in the confined zone at or near the moving element, diffusion is not a likely capture mechanism. Diffusion is seen more often in designs that create a spray and sustain that high surface area spray for an extended time. Likewise, many of these devices use centrifugal force applied close to the moving element to separate the gas and liquid; therefore, gas absorption is somewhat limited as the liquid surface area per unit volume is decreased during this separation.

11.5 Design Basics

Most mechanically aided scrubbers are proprietary designs that have been refined over the past few decades.

In general, the gas inlets and outlets are sized based on the conveying velocities that these devices need to successfully control dust at high loadings. Gas inlet velocities of 45–60 ft/s are not uncommon. The ducting to the moving device must be carefully designed to load the scrubber uniformly. Obviously, imbalance can be a problem with any moving device; therefore, the designers take care to allow uniform loading and to clean surfaces upon which dust may build.

The total horsepower input of these devices typically includes the energy required to move the gas, liquid, and dust. Also, a pump or source of pressurized liquid is needed.

11.6 Operating Suggestions

The vendors of these devices have accumulated a wealth of experience in a wide variety of applications. It is best to contact them regarding specific requirements.

Because many of these designs use spray nozzles at some point, the use of a solids separation device (strainer) is often required but not a standard part of the scope of equipment supply. The drag chain shown in Figure 11.1 helps remove large solids before they reach an injurious concentration level. In addition, external strainers or liquid cyclones are sometimes suggested to remove the captured particulate.

Given their simple designs, mechanically aided scrubbers often use a simple wire type overflow for level control. Open impeller pumps are often seen on these designs, given their use on generally high-solids applications. Spare pump wetted parts such as impellers, seals, shafts, and shaft sleeves are recommended.

Because the moving element is often sprayed, vibration detection devices and amperage meters can be used to monitor that important element. Increases in fan amperage can offer a warning of wheel solids buildup. A differential pressure gauge is often used to monitor the scrubber pressure drop. In keeping with the simplicity of the design, little if any additional instrumentation is used.

When the scrubbers are shut down for inspection or service, particular attention should be paid to corrosion or wear on the rotating element. Firms that use multiple mechanically aided scrubbers (such as foundries for sand dust separation) often have spare rotating elements on hand.

12

Packed Towers

12.1 Device Type

Packed towers are gas absorption devices that utilize internal media of a variety of types to enhance the mass transfer of gases into an absorbing liquid. Please also see filament/mesh scrubbers (Chapter 9) which share many of the same design and use characteristics of the packed tower.

12.2 Typical Applications and Uses

For both air pollution control and recovery of process gases, packed towers are one of the most common mass transfer devices in current use.

They are used for control of soluble gases such as halide acids (such as HF and HCl) and to remove soluble organic compounds such as alcohols and aldehydes. When the scrubbing solution is charged with an oxidant such as sodium hypochlorite, they are used to control sulfide odors from wastewater treatment facilities and rendering plants. They are used to absorb and concentrate acids for recovery. When gases and aerosols are both present, the packed tower is frequently used ahead of aerosol collectors such as fiberbeds and wet electrostatic precipitators.

Packed towers are also used as gas coolers and condensers (Please see Chapter 22 on energy recovery). They sometimes are used after a hot gas quencher to act as a gas cooler. Some are fitted with ceramic packing that can resist temperature extremes. When fitted ahead of a Venturi scrubber to function as a water vapor condenser/absorber, the packed tower becomes a critical part of a flux force condensation system for particulate control. The tower in this case acts as both an acid gas absorber and a direct contact vapor condenser.

They are also used after Venturi scrubbers on medical waste incinerators to control acid gases such as HCl.

To control the combined vent gases from semiconductor manufacturing, large, packed towers are used. Called *house scrubbers*, they clean the small concentrations of acid gases usually using pH control and neutralization with caustic. In contrast, the same industry uses small, packed towers at specific tools in a point-of-use configuration. The point-of-use scrubbers are designed to treat the specific emission source and often vent into a combined ventilation system, eventually leading to a house scrubber. The emissions are effectively double scrubbed before the carrying gas is released to atmosphere.

Pulp and paper mills often use packed towers for bleach plant applications to control chlorine and chlorine dioxide where fibers or chemical scaling is minimal. Fluidized bed-type scrubbers are used in cases where fibers or scaling are known challenges.

12.3 Operating Principles

As mentioned in Chapter 1, absorbers function by extending the surface of a solvent (usually water) so that the mass transfer of a gas into that solvent is enhanced. The mass transfer of a gas into the liquid is limited by the gas/liquid interface conditions. Only a certain mass of gas can move into the liquid per unit *area*. Once into the liquid, only a certain amount of dissolved gas can remain, per unit *volume*. Therefore, to effectively remove the gas, one must have sufficient liquid surface area and an adequate volume of liquid.

The packing (or medium) in a packed tower provides the liquid-extending function to increase its area. The liquid inlet system provides the adequate volume. By selecting the proper type and quantity of media, the conditions can be created for optimum mass transfer. The result is a tower containing the design amount of media (or an excess) irrigated by the design amount of liquid (or an excess). If the gas flows vertically, the tower may contain just a few feet of this medium, or over 50 ft of medium, depending upon the absorption characteristics of the contaminant and the neutralizing capability of the liquid. Towers may also be required in series to reach the desired gas outlet conditions.

Packed towers are essentially probability machines. The individual contaminant gas molecule is in contact with the descending liquid for only a fraction of a second. By increasing the number of chances of such random contacts through increasing the height of the packed bed, the chances that the molecule will be absorbed are increased. If you do not absorb it now, you might absorb it later. Also, it takes time for the gas to diffuse to the liquid surface. If one gives such diffusion more time by letting the gas move slowly through a long-contact bed, one increases the chance of successful absorption.

The standard vertical (counterflow) packed tower has the components shown in Figure 12.1. The vessel contains a grid that supports the packing medium. The medium is irrigated from above by a liquid distribution device (usually a spray header or headers). The liquid hits the medium and

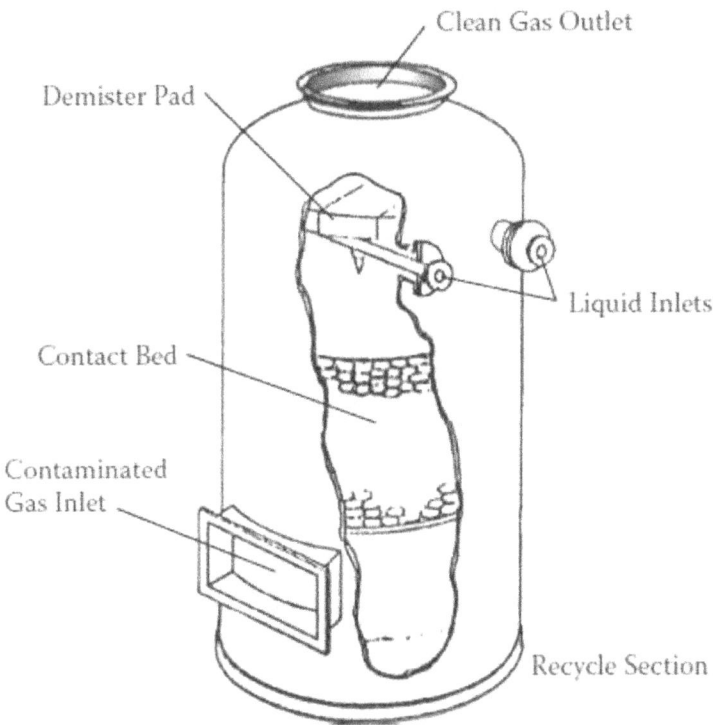

FIGURE 12.1
Vertical counterflow packed tower components (Bionomic Industries, Inc.).

high-surface area liquid films and/or drip points are formed as the liquid flows over and through the medium. The gas, flowing in the opposite direction as the liquid, is caused to take a tortuous path through the medium, thus bringing the gas close to the absorbing liquid. The gas contacts the liquid surface and, if the liquid is not saturated with the contaminant, is absorbed. If some contaminant is already present in the liquid, not all the contaminant gas will be absorbed. Therefore, a large volume of packing is often used so that, particularly at the top of the packed tower, the scrubbing liquid can absorb and retain the gas. If not, the removal efficiency of the packed tower will be reduced.

A crossflow arrangement (Figure 12.2) is similar except that some of the gas and liquid move concurrently and that the liquid is rejected downward along the entire vessel path length. For gases that are absorbed and react with dissolved compounds, the crossflow and counterflow towers behave similarly. If the gas does not react with chemicals in the liquid, the crossflow tower can demonstrate a reduced efficiency since the liquid is carried, with its dissolved gas cargo, toward the gas discharge point, creating a vapor pressure condition that favors the gas. This means that the liquid may not have

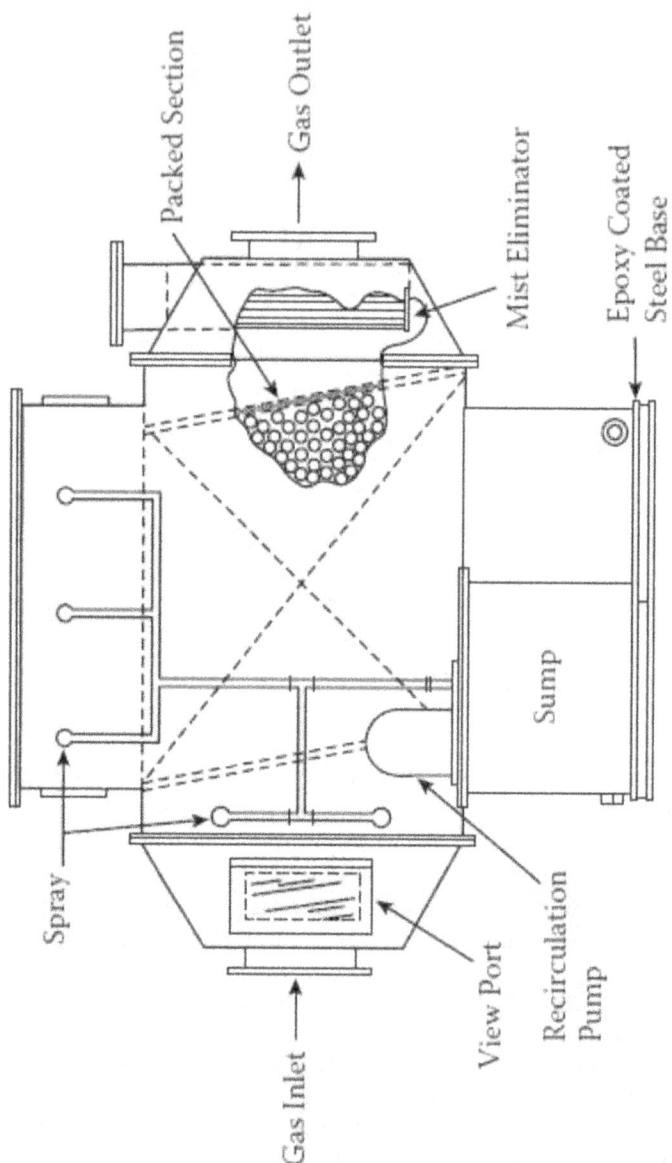

FIGURE 12.2
Crossflow packed tower components (Bionomic Industries, Inc.).

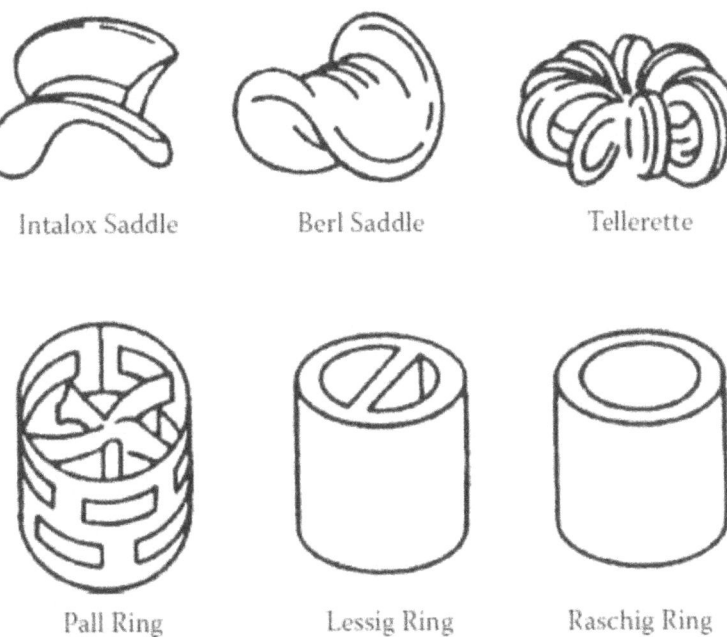

Intalox Saddle   Berl Saddle   Tellerette

Pall Ring   Lessig Ring   Raschig Ring

FIGURE 12.3
Dumped-type packing media (Bionomic Industries, Inc.).

the same absorption capacity in the crossflow design as in the counterflow design when no liquid phase reaction occurs.

There are hundreds of types of packed tower packing material that forms the packed bed. Figure 12.3 shows a variety of basic types of dumped-type packing media. These media may be made from thermoplastic material such as polypropylene, metals such as stainless steel or corrosion resistant alloys, or even in the form of cast ceramics. Figure 12.4 shows media offered by RVT Process Equipment, Inc., and Figure 12.5 shows media designed and supplied by Lantec Products, two of the leading domestic suppliers of this type of media.

You can see by the designs that certain configurations produce large surface films and others have small holes or openings that form numerous drip points. In general, where scaling can occur, packing with large openings that produce drip passages rather than film surfaces are used because scaling is a surface phenomenon. The various vendors seek to combine a balance between mass transfer enhancement and plugging and scaling resistance. The resulting packing must be structurally sound as well because the material rests on and is supported by the medium beneath it. In a more subtle manner, the packing must resist side-to-side motion under the influence of gas or liquid flow. If the packing moves around easily, valleys or mounds of packing can form in the tower, upsetting its performance.

FIGURE 12.4
RVT packed tower hiflow media (RVT Process Equipment, Inc.).

FIGURE 12.5
Lantec packed tower media (Lantec Products, Inc.).

FIGURE 12.6
Structured ceramic packing (Lantec Products, Inc.).

Media can also be in the form of shaped and/or perforated panels. This is called *structured packing* because the media are structurally self-supporting. Figure 12.6 shows a type of structural packing. The plastic versions are cousins to cooling tower fill, and many look like corrugated plastic panels. Other fill material is made of woven mesh, much like the mesh used in a mesh pad droplet eliminator. This type of medium is used in distillation columns and applications, in general, where no solids are present. If solids are present, the medium can act as a liquid filter and plug.

12.4 Primary Mechanisms Used

Gas absorption in a counterflow (vertical gas flow) packed tower is dictated by the equilibrium conditions between the contaminant gas and the absorbing liquid. The overall controlling mechanisms are ruled by the solubility of the gas in the liquid and by any reactions that may be caused to occur in the liquid with a reacting chemical. If the gas reacts with a chemical forming a lower vapor pressure compound, the equilibrium shift favors further absorption. If the absorbed gas builds up in the liquid, the equilibrium shifts to inhibit subsequent absorption.

Diffusion is used to move the gas to the liquid surface. At or near the liquid surface, phoretic forces such as thermophoresis or diffusiophoresis may be in play.

In essence, however, packed towers are equilibrium and probability machines. The overall gas/liquid equilibrium controls the design of the tower. Because the gas is absorbed at the liquid surface, the more liquid-to-gas interactions that can be caused to occur, the greater the probability of absorption. The more difficult the absorption, therefore, the greater the medium depth. This increases the number of contact possibilities, thus increasing the likelihood that a contact will be successful, and the gas will be absorbed.

12.5 Design Basics

The contaminant solubility, vapor pressure characteristics, and the scrubbing liquid's capacity for that contaminant control the actual amount of packing needed in a packed tower. Packing selection is covered in detail in books specifically devoted to mass transfer (see Appendix A) and is beyond the scope of this book.

A method has been developed to compare various packing types. This parameter is called the packing factor, and you will see specific packing factors published for various packing types. Most packing vendors, however, will provide for you the estimated packing quantity for their specific packing after you submit the gas flow and scrubbing liquid characteristics to them. Some will even design towers for you. It is advised, however, that you solicit the assistance of an experienced packed tower vendor before committing to a tower selection. These devices are more complicated than they appear to be on the surface.

12.5.1 Counterflow

Gas inlet velocities are usually 40–55 ft/s in packed towers. The inlet velocity is usually dictated by common ventilation system design practice. In vertical counterflow tower designs, the vessel gas velocity is 3–8 ft/s. The upper limit is dictated by the flooding characteristics of the packing.

Any packing can flood. Flooding occurs when the gas kinetic energy is sufficient to hold up all the scrubbing liquid. The liquid spreads out across the tower seeking some means to drain but cannot. The pressure drop of the tower starts to swing or surge and the hydraulics become unstable. For most gas absorption problems at near-ambient conditions, at approximately 8 ft/s, the tower might flood. Packing vendors perform tests on their packing and determine flooding velocities and gas mass flow rates for their various packing types. The designer sizes the vessel to stay below that flooding point.

Ironically, most mass transfer operations reach their peak efficiency just before flooding occurs. Mechanically, however, the stability of the tower decreases as one approaches flooding. A compromise is needed. Most towers are designed for less than 80% of predicted flooding.

To support the packing, flat or curved injection type grids are used. Figure 12.7 is a rendering of an injection type grid. The curved surfaces allow the ascending gas to be injected into the packing not on one plane but over a deep zone. The gas can enter the packing at an angle, thereby allowing the liquid to drain more readily.

If dumped-type packing is used, hold-down grids are often used to hold the packing within the required absorption zone.

The liquid itself is distributed by spray headers as shown in Figure 12.8 or by distribution weirs as shown in Figure 12.9. Care is taken with spray-type distributors to make certain that the spray patterns overlap but do not impact the vessel wall excessively. If the liquid hits the wall, it forms sheets of liquid

FIGURE 12.7
Injection type packing support (RVT Process Equipment, Inc.).

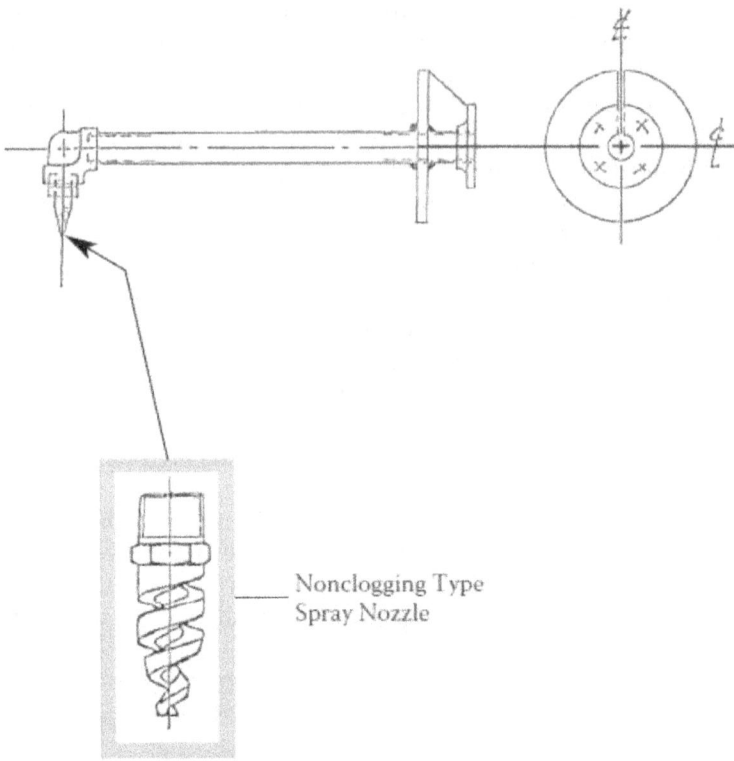

Nonclogging Type
Spray Nozzle

FIGURE 12.8
Retractable liquid distribution headers (Bionomic Industries, Inc.).

FIGURE 12.9
Liquid distribution weir boxes (RVT Process Equipment, Inc.).

on the wall that are largely ineffective in absorption because it only attains the area of the vessel wall itself. Many vendors of packed tower internals offer proprietary liquid distributors. These designs often have their roots in distillation towers and are highly engineered (and tested) to produce a uniform liquid loading. If spray headers are used, the liquid velocity is 4–8 ft/s. Free-flow fittings to distributor trays are in the 3–4 ft/s range, sometimes lower.

Packing is usually irrigated at a minimum of about 6–8 gallons per minute (gpm) of liquid per square foot of packing. It is not unusual to irrigate at over 20 gpm per square foot to make certain that all the packing is wetted. If the packing is not fully wetted, the performance of the scrubber will be reduced.

If a packing depth of more than about 10 feet is required, redistributors or rosettes are used to pull liquid from the wall toward the center. The gas velocity through the packing causes the pushing of the liquid toward the wall. The velocity tends to be slightly higher at the center than the wall, thus the liquid is ejected toward the wall. The rosettes act as baffles to direct the liquid back toward the vessel center, thereby keeping all the packing wetted.

The upper surface of the packing and its liquid distributor generate residual droplets that are controlled by a mist eliminator. Mesh pads are often used when the gas stream is clean (no solids) or chevrons when some particulate may be present. Mesh pads require a gas velocity of about 10 ft/s or less. Chevrons permit higher gas velocity (10–12 ft/sec), but this would require a change in vessel diameter. As a result, the packed tower vessel is usually designed for about 8 ft/s or less. If a chevron is used, its face area is reduced using a blank-off plate.

12.5.2 Crossflow

Gas inlet velocities are in accordance with the counterflow designs. Because the gas flows side to side in the crossflow design, the liquid is draining out of the gas stream, so the packing resists flooding. As a result, the crossflow orientation can run at higher gas velocities.

The box velocity is usually 5–10 ft/s. The liquid loading can be higher than that used in the counterflow packed tower. This higher liquid capacity can be an advantage where the gas is only slightly soluble in water (you can use more water).

The droplet eliminator in the crossflow also rejects liquid out of the air stream, but it ejects it out and down, rather than back into the gas stream. If a chevron is chosen, it can operate at 12–15 ft/s. If a mesh pad is used, velocities of 8–12 ft/s and sometimes higher are possible. The mesh pads are often inclined to enhance draining along its element (the gas drains at an angle given the gas velocity pushing the liquid to the side). Either of these devices is often mounted in a containing box with a flanged service cover.

With crossflow designs, a reduction in efficiency can occur if the gas short-circuits over the top of the packing. To prevent this, vendors use baffles or extend the packing up into the box area above the packing. Others place a layer of mesh pad above the packing to offer greater resistance to gas flow. Still others use two or three different packing sizes so that the gas is pushed lower in the tower. The more resistant packing is placed at the top, near the irrigation headers.

The liquid is distributed much like in a counterflow packed tower at similar velocities. The liquid header pressure need not be high since the liquid nozzles are within a foot or so of the packing. Pressures of less than 10 psig are used firing full cone nozzles. Some designs use pipes with holes in them, thereby eliminating the nozzles.

12.6 Operating Suggestions

Packed towers, particularly vertical gas flow types, need to be installed vertically. A plumb line is often used to help set the verticality of the unit.

If the tower is made from fiberglass-reinforced plastic and is installed on a concrete pad, roofing felt (tar paper) is placed under the tower to compensate for pad irregularities. If the towers are installed on steel plates, roofing felt is also used to allow the plastic packed tower to expand and contract with minimum chafing on the plate.

Always plan the surrounding area for packing removal and installation. Sometimes height constraints eliminate the possibility of using access doors above and below the packed bed. In this case, the whole top of the tower may have to be flanged and bolted for removal.

On towers equipped with liquid headers, to access the nozzles, retractable, flanged headers are suggested. If these headers are plastic and are less than 3-ft long, they can be cantilevered. If they are longer, they typically need to be extended fully across the tower diameter and be retained in a socket or similar support. Why? When the header is pressurized, the reaction force of the liquid ejecting from the nozzle tends to push the header upward. When unpressurized, the header tends to sag. The end support reduces both effects.

If caustic is used in packed towers, it should be thoroughly mixed. One way to do this is to inject it into the liquid recirculation circuit ahead of an in-line mixer. Another way is to take part of the recycle liquid and divert it into a submerged sparger located in the scrubber sump. The caustic is injected into this sparger and is thoroughly mixed in the sump.

Using a differential pressure gauge or transmitter monitoring the bed pressure drop can reveal the condition of the packed tower. All things being equal, if the pressure drop rises, the bed may be plugging.

Acid washing a scaled-up packed bed can be difficult. It is much like trying to clean both sides of an umbrella by sending liquid down on it. You can possibly clean the upward-facing side, but what about the underside?

The only truly effective method is to totally flood the tower with descaling chemical (usually acid for carbonate scale and caustic for silicate scale). The other method is to remove the packing and wash or replace it. The latter is the most common method.

Care should be taken in packed towers using spray nozzles to provide strainers to remove or trap solids that could plug the nozzles. Some vendors use removable perforated plates that trap solids. Others use single or duplex basket strainers.

Packed towers offer efficient control of soluble gases in environments in which solids plugging, either by solids in the gas stream or by-products of the gas/liquid reaction, is minimal.

13

Settling Chambers

13.1 Device Type

One of the simplest (and oldest) air pollution control devices is the *settling chamber*. These are also sometimes called *knock-out boxes* or *drop-out boxes*. The equipment is in the form of a large chamber, which allows reduction of the gas velocity to a point where the particulate it carries simply drops out.

Today, settling chambers are used for coarse removal of large particulate in advance of higher efficiency particulate control equipment.

They are rarely, if ever, used as the final gas cleaning device.

13.2 Typical Applications and Uses

Settling chambers are primarily used to reduce the loading of particulate from sources such as kilns, calciners, and mills or grinders that inherently produce high-particulate concentrations. If the particulate is valuable in a dry form, the settling chamber usually is designed to settle out the smallest size particle that can economically be separated. If the product is not valuable or further downstream particulate separation is to be used (such as a cyclone, scrubber, or fabric filter collector), the chamber is usually sized to afford some basic separation at low cost.

They are often followed by product recovery cyclones which are, in turn, followed by collectors designed for high-efficiency collection of the fine particulate that pass through the upstream devices.

13.3 Operating Principles

A settling chamber operates on the principle that if you slow a gas stream down sufficiently, the solid particulate contained within that gas stream will settle out by gravity. In general, the larger the particle, the faster the settling rate. In addition, larger particles will settle out faster in each moving gas stream than smaller particles.

The settling velocity for particulate was explored extensively in the mid-1800s by a scientist named Stokes. His equation for the terminal settling velocity of particulate is used to this day. It is called Stokes' law:

$$V_g = (D^2(d_p - d_g)g)/18v$$

where:

V_g = terminal settling velocity (ft/s)
D = particle diameter in feet
d_p = density of particle, lbsm/ft^3
d_g = density of gas, lbsm/ft^3 (pounds [mass])
g = acceleration of gravity, ft/s^2
v = gas viscosity, lbm/ft/s

The settling relationship is accurately applied only for particles of about 2 μm and greater aerodynamic diameter. Usually, for calculations involving air at ambient conditions, the density of the gas is ignored because it is minor when compared with the particle density.

What this equation shows is that the greater the particle diameter and density, the higher the particle's settling velocity. Resisting this settling, the higher the viscosity of the gas, the lower the particle's terminal settling velocity.

Settling chambers are therefore designed to allow the mean gas stream velocity to slow down to a point at or below the target particle's settling velocity so that the particle drops out within the confines of the chamber. Because the particle settles at a given rate (i.e., distance per unit time) as predicted by Stokes' law, the chamber must be sufficiently long to allow this settling to be completed before the gas reaches the device's gas outlet. Settling chambers are therefore large in cross-sectional area, to slow the gas stream down, and long, to allow sufficient time for settling.

What about particles under approximately 2 μm diameter? Unfortunately, these particles (about $\frac{1}{25}$th the diameter of a human hair) are so small that they are influenced greatly by surrounding gas molecules and do not follow Stokes' law. They do not even follow a trajectory as such. They are buoyed and buffeted by surrounding gas molecules. A correction for Stokes' law was derived by a researcher named Cunningham. Thus, we have Cunningham's correction factor for non-Stokes-sized particles. Sometimes called a slip-correction factor, it is a multiplier applied to Stokes' equation to adjust for the particle size and its actions below 2 μm aerodynamic diameter.

Experience has shown that settling chambers are of practical value only for reducing the loading (concentration) of large (above 100 μm aerodynamic diameter) particulate and possibly for the recovery of large, valuable product. They are used in various design configurations on devices such as kilns and calciners, waste solid fuel boilers, or similar devices. They are almost invariably followed by more efficient gas cleaning equipment.

Dust chambers are often used at the feed end of lightweight aggregate kilns. Similar knock-out chambers are used on mineral lime, cement, and lime sludge kilns and sometimes on dryers. Large particulate that is air-conveyed out of the rotating portion of the kiln/dryer is encouraged to drop out in the knock-out chamber and be recovered. Sometimes, a vertical baffle is used in the chamber to direct the gas stream in a pattern that makes the gas perform a 90° or even 180° turn to enhance separation. The larger particulate cannot make this turn and therefore drops out.

13.4 Primary Mechanisms Used

The primary mechanism used is the drag force applied on the particle by the viscosity of the carrier gas. As the gas stream slows down, the influence of the viscous force of the gas on the particle is reduced and the particle begins to settle by primarily gravitational forces.

13.5 Design Basics

Settling chamber design is predicated on the particle size, its density, the gas viscosity and velocity, and space considerations. An infinitely large settling chamber would, in theory at least, settle out all particulate. Economics, however, limits the size of the chamber. Stokes, in turn, limits the size of the particle that can be economically separated.

If the chamber is used for valuable product recovery, the smallest particle that would be worthwhile collecting dry is the common target. The design focus then needs to answer the question, Is there enough space? An iterative design then follows. As mentioned earlier, Stokes' law defines the settling velocity, and the velocity dictates the size of the equipment. This usually results in a design particle size more than 50–100 μm, otherwise the chamber becomes excessively large. If the 50–100 μm particle is not worth collecting, the designer would size the chamber to capture much larger particles, thereby at least economically lowering the loading of particles requiring further control but letting the smaller particles pass through.

Chamber (or can) velocities of 5–7 ft/s or lower are common. Baffles can sometimes be used to provide beneficial changes of direction if the particles do not stick to the baffles. Curtains of chains can be used to in effect divert the gas flow but allow some measure of self-cleaning. Given the low gas velocity, the pressure drop is usually under 1-inch water column.

Figure 13.1 from Fan Engineering (Howden North America, Inc.) shows a general diagram of a crossflow settling chamber. Note the hoppers used

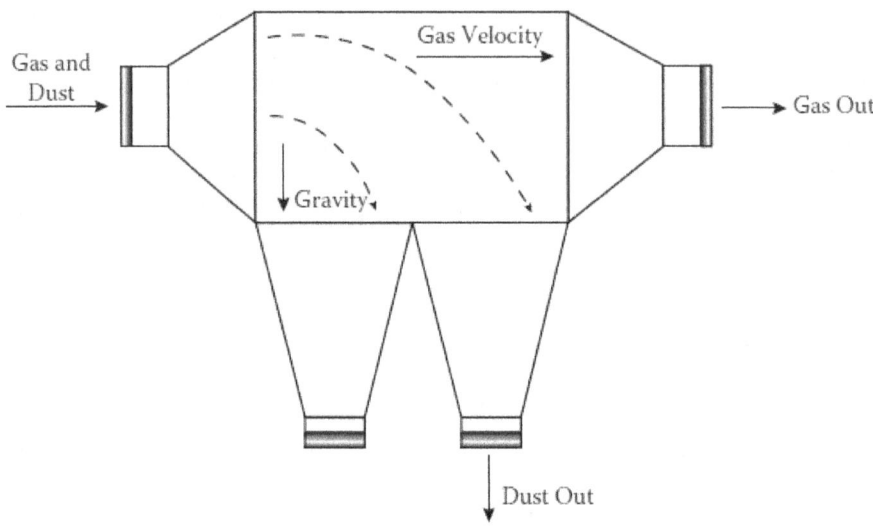

FIGURE 13.1
Settling chamber.

to remove the collected solids. Gas flow is left to right. The vector diagram depicts the primary forces on the particle, which influences the trajectory and, therefore, the length of the settling chamber.

Even given a dispersion of particulate above 100 μm, the efficiency of a settling chamber is quite low. Typically, only 25%–50% of the particulate of that range or larger drops out. Settling chambers are often, therefore, called *rock boxes* in the industry because they remove only the "boulders." In doing so, however, they can serve a valuable purpose in reducing the total loading of particulate that must be removed by downstream devices.

13.6 Operating/Application Suggestions

Most often, the designer of the process equipment includes a settling chamber in his design as an integral part of the device. The settling function may be just a minor one. The primary purpose may be to allow material to be fed into the device or to allow for seals and so on to function properly. It is therefore best to use the design provided by or recommended by the process equipment vendor.

If a settling chamber is used, care should be taken to design a suitable solids discharge system so that the particulate does not build up to a point where it entrains into the gas stream. Access doors should be provided for service access and cleaning. If the gas stream contains acids and its temperature and humidity pass through the acid dewpoint, the chamber should be suitably insulated and even heated to reduce corrosive effects.

The structural support for a settling chamber should be sufficient to support the filled weight of the device. This can be a significant factor, since these devices are inherently large.

Settling chambers should not be used where the particulate is sticky or can bridge or build up. In those cases, quite the opposite design is used. The ductwork is sized to be above the conveying velocity of the target particulate, and that velocity is maintained until the particulate reaches a suitable gas cleaning device.

14

Spray Towers/Scrubbers

14.1 Device Type

Spray tower scrubbers use spray nozzles to extend the surface area of the scrubbing liquid to enhance mass transfer of contaminant gas(es) into the liquid. They are primarily used for gas absorption.

Spray scrubbers include designs that use spray nozzles (hydraulically or air or steam atomized) to absorb gases and control larger diameter (+10 µ) particulate.

14.2 Typical Applications and Uses

Spray tower scrubbers are often used on wet flue gas desulfurization (FGD) systems at public and industrial power generation facilities. These FGD systems use lime or limestone slurries as the scrubbing liquid. Their open vessel design is an advantage where plugging or scaling may occur. The simplicity of the design makes them a lower-cost alternative for high gas volume scrubbing applications (over 100,000 acfm).

They are also used as part of quenching and gas conditioning systems wherein the gas must be brought to saturation or near saturation with water.

Most spray towers are countercurrent in design wherein the gas flows vertically upward and the liquid falls downward through the ascending gases. Some units, used for odor control, are horizontally oriented using a multiplicity of concurrent spray sections in series. Often, a series of baffles is incorporated into the scrubber vessel to change the direction of the gas stream and to provide intimate contact of the spray and the contaminant gas.

Some spray tower designs have been modified to act as direct contact condensers in applications wherein packed devices may plug with solids. Though not as mechanically efficient in that application as pure countercurrent designs (such as a packed tower), the ability to resist plugging and therefore provide greater on-line availability can make the spray tower condenser attractive.

Spray scrubbers cover a wider variety of designs. These vary from devices as simple as a spray header in a duct to cyclonic-type devices (often called preformed spray scrubbers).

14.3 Operating Principles

A common characteristic of this type of scrubber is the use of spray nozzles to extend the liquid surface and produce target droplets.

At least one spray zone is produced in a spray tower using at least one spray nozzle in a containing vessel. In practice, however, most spray towers use multiple spray zones to achieve the required gas cleaning efficiency. Figure 14.1 shows a sectional view of a spray tower. The gas inlet is typically

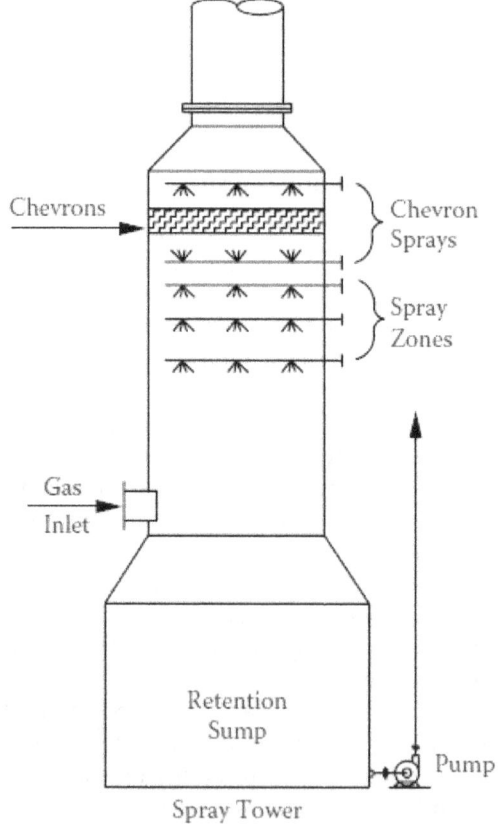

FIGURE 14.1
Spray tower sectional view.

horizontally oriented into the containing vessel. A multiplicity of spray zones is used, each containing an array of nozzles. In FGD applications, these nozzles are wear-resistant designs (such as silicon carbide) since the scrubbing liquid is an abrasive slurry of lime or limestone.

The hydraulic pressure applied to the liquid acts as stored energy. When this pressurized liquid flashes from the spray nozzle, the energy stored is expended in producing a spray. The high relative velocity between the liquid and surrounding gas causes a shearing action that breaks the liquid into tiny droplets. The net effect is that the liquid surface area increases so that the contaminant gas or gases can be more readily absorbed.

After the spray is produced, the contaminant gas is absorbed through the liquid film. If a reactive chemical is contained in the droplet, the contaminant will react, forming a by-product (usually a salt) of lower vapor pressure. Therefore, the contaminant remains in the droplet.

Most droplets fall by gravity in counterflow designs to the sump. Quite often, the scrubber is mounted directly over the sump to facilitate this separation. A small portion of the spray goes overhead with the gas. This droplet dispersion is controlled using chevron-type droplet eliminator(s) in the case of a gas stream containing particulate, or mesh pads if the gas stream is low in or devoid of particulate. The chevron droplet eliminators are often sprayed constantly from below and on a timer basis from above for cleaning purposes.

Figure 14.2 shows a common application where a utility FGD spray tower is installed after a dry precipitator used for particulate control.

FIGURE 14.2
Utility FGD system (Babcock & Wilcox Co.).

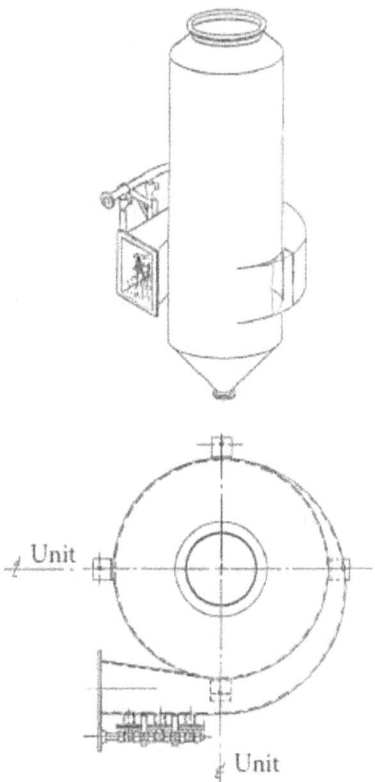

FIGURE 14.3
Preformed spray scrubber (Bionomic Industries, Inc.).

With the preformed spray scrubber, the spray nozzles are generally installed in the gas inlet area of essentially a cyclonic separator. The spray dispersion is very intense and dense in the inlet zone. The gas is accelerated as the gas approaches the tangent point of the separator vessel. This action enhances particulate capture. The droplets are then spun from the gas stream using centrifugal force. Figure 14.3 shows a sketch of a preformed spray scrubber in elevation and plan view. Note how the gas inlet curves around the cylindrical separator vessel. This curved portion is called an *involute* and may extend from 90° to 270° of vessel circumference. Note also that the sprays are mounted on individual headers on the involute for simplified access. These headers usually are connected to a distribution pipe by hoses and are isolated by valves so that individual headers may be removed for servicing.

A preformed spray scrubber was used on the superphosphate fertilizer multistage scrubber application referenced in previous chapters. It forms the base of the stack as shown in the center of Figure 14.4. It was used to remove residual fluoride compounds and to concentrate the fluosilicic acid prior to

FIGURE 14.4
Preformed spray scrubber on superphosphate dryer (Bionomic Industries, Inc.).

the solution being sent to the filament/mesh pad scrubber (to the left) for further concentration. The fluosilicic acid recovery tanks are to the right in the picture.

14.4 Primary Mechanisms Used

The primary scrubbing mechanism used in a spray tower is absorption. To some extent, diffusion is in play as the contaminant gas moves toward the droplet surface. The droplets themselves can remove some particulate by impaction; however, the relative velocity between the gas and liquid is low (usually below 20–40 ft/s), so impaction is minor.

Spray scrubbers using cyclonic action do apply impaction and interception forces to the gas stream and therefore exhibit higher particulate removal rates.

14.5 Design Basics

Gas inlet velocities of spray towers are in the range of 50 to 60 ft/s as is common with other wet scrubbing systems. Sometimes, for gas distribution purposes, the gas is conveyed to the scrubber at this velocity to keep particulate entrained but is reduced to 40–45 ft/s at the scrubber itself.

Countercurrent spray towers normally operate at vertical gas velocities of 8–10 ft/s; however, in recent years, efforts have been made to operate them at up to 15 ft/s. At approximately 15–16 ft/s gas velocity, the descending spray tends to be held up or fluidize. At this point, the spray tower begins to transition to a fluidized bed scrubber. The spray nozzle method of liquid injection becomes of diminishing importance as the gas velocity rises since the spray is created by the ascending gas at these speeds.

The chevron zones of these designs usually use a face (open vessel) velocity of approximately 10–12 ft/s. Interface trays, much like weeping sieve trays, are used in some designs to suppress liquid carryover and isolate the dilute wash water spray that is applied to clean the chevrons.

If a top-mounted stack is used, the gas outlet velocity will be often under 45 ft/s to reduce the chance of entrainment. Speeds of 35–45 ft/s are common.

For FGD systems, the pH (and sometimes the density) of the scrubbing solution is controlled to operate within a window bounded by efficiency and scaling. For limestone slurry scrubbing, this results in a pH range of approximately 5.6–6.5 in the slurry and 5.4–6.2 in the sump. For lime, the slurry is approximately pH 7–8 going into the absorber and 5–5.5 in the sump.

Nozzle pressures of 30–60 psig are used, depending on the application.

Given that the liquid surface area of a spray decreases as the distance from the nozzle increases, high liquid-to-gas (L/G) ratios are used, that is, using multiple nozzles, to maintain the net surface area at a sufficiently high level. As a result, it is not uncommon to see L/G ratios of 50–100 gpm/1000 acfm treated being used in these designs. The pumping cost, therefore, becomes a significant design factor.

Given the open area of the vessel, however, the gas side pressure drop is quite low. Spray towers operate at pressure drops of only 1–3 inches water column. This keeps the fan horsepower low. This factor is significant for high gas volume applications.

Spray towers of more than 30 ft diameter have been built. The simple vessel design allows these large-diameter vessels to be made. For high gas volumes, multiple towers are used in parallel. On utility boiler systems, redundancy is often built-in by having the capability to switch between operating and standby vessels using suitable isolation dampers.

14.6 Operating Suggestions

Over the past 40 years, operators of spray towers have developed specific methods for the best operation of these devices.

Some basic techniques include separation of solids that would be sufficiently large to plug the spray nozzles. Settling tanks and liquid cyclones are often used to separate the large solids. The nozzles themselves are designed for high solids throughput and wear resistance. Often of full cone design, the nozzles are arranged in patterns that cover the vessel but reduce zones where agglomeration of droplets (resulting in an undesirable reduction of surface area per unit volume) can occur. Multiple spray levels increase the probability that all zones are covered.

Once absorption occurs, the chemical reaction kinetics in the liquid may be slow. In FGD systems, the scrubbing liquid is often impounded in an agitated tank to allow crystal formation and settling. Residual crystals can recycle to help scour the scrubber interior and reduce hard scaling. Sometimes, chemical additives (such as adipic acid) are administered to improve the scrubbing performance. Oxygen is sometimes injected to oxidize the sulfite component of the scrubbing solution to sulfate so that the sulfate may be more easily settled and removed.

For materials of construction, the vessels are often mild steel with rubber lining for utility FGD application. If chlorides are present, alloys such as 904L, AL6XN, C-22, or C-276 are used.

Chevrons in many FGD designs are installed in stages given the high droplet loading. A coarse stage of widely spaced blades is used followed by more narrowly spaced chevrons in either vertical flow or horizontal flow configuration. Figure 14.5 shows a chevron set using multiple design configurations to arrest the spray.

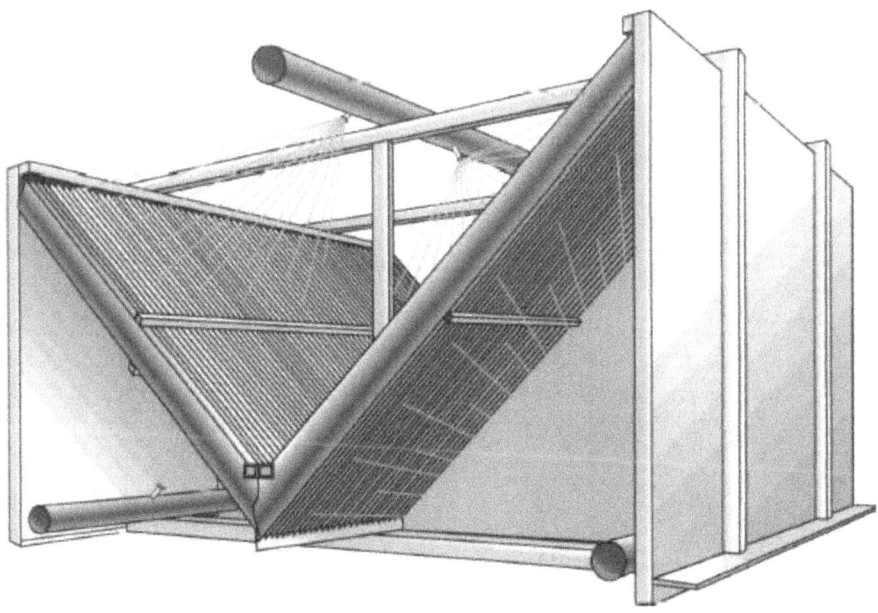

FIGURE 14.5
Multiple chevron stages (Munters Corp.).

Spray scrubbers have been made in a wide variety of materials, from carbon steel, to rubber-lined steel, to fiberglass-reinforced plastic, to exotic alloys. Some designs have even received food-grade interior polishing to handle explosive-type material. Preformed spray scrubbers usually are equipped with retractable spray headers and individual shut-off valves for nozzle servicing. Obviously, one must plan for sufficient pull space to remove such headers.

If large solids are anticipated that could plug the nozzles, strainers on the recirculation loop should be used. It is also advised to locate any vessel access door such that the worker can gain entrance to the scrubber easily. A common location is directly over the separator inlet duct.

Preformed spray scrubbers perform like Venturi scrubbers operating at 6–10 inches water column pressure drop. This means they are best suited for the control of particulate above 10 µm aerodynamic diameter. For gas absorption, an inlet spray type unit can achieve about 0.8–1.5 transfer units of separation. Ones with wall-mounted sprays can often achieve higher mass transfer rates but are more likely to entrain droplets.

Spray towers and spray scrubbers are popular devices for use in gas absorption applications and, in the case of preformed spray scrubbers, for particulate control on particles more than 10 µm diameter.

15

Thermal Nitrogen Oxide (NO$_x$) Control

Joseph Colannino

Colannino Associates, Oceanside, California

15.1 Device Type

The control of nitrogen oxides (NO$_x$) using thermal methods encompasses a variety of devices. This chapter focuses on NO$_x$ and its control using combustion modifications, post-combustion thermal and catalytic methods, and combinations thereof.

15.2 Typical Applications and Uses: Combustion Sources

Various combustion sources produce NO$_x$. *Boilers* use a burner to combust the fuel and release heat. The heat boils water and generates steam. Larger boilers usually contain the water and steam inside tubes (water-tube boilers) surrounding a firebox. Some smaller boilers have a combustion tunnel surrounded by water (fire-tube boilers). The water-tube boiler has an analog in the petroleum refinery—the *process heater*.

The process heater is used to heat or transform a process fluid, for example, crude oil. Analogous to the water-tube boiler, the process fluid is pumped through tubes surrounding a firebox. Most boilers are heated with burners in the horizontal direction. Process heaters are often fired with the burners in the floor. However, some process heaters are wall fired, and some specialty reactors such as reformers are down fired from the roof. Process heaters may be tall, round floor-fired units (known as vertical cylindrical heaters) or rectangular units known as cabin type, which are often floor fired but may also be wall fired. Some specialty heaters, such as ethylene cracking furnaces and reformers, use heat to chemically transform the process fluid.

Gas turbines and reciprocating engines transform heat into mechanical motion. Hazardous waste incinerators use high temperatures to destroy waste products. All conventional combustion processes form NO$_x$.

15.3 Operating Principles

NO_x are criteria pollutants as classified by the Environmental Protection Agency (EPA). Accordingly, the EPA has established National Ambient Air Quality Standards (NAAQS). Local air quality districts translate the NAAQS into local regulations for various combustion sources. These regulations vary widely from region to region. The purpose of this chapter is to show how NO_x is formed and to discuss some methods for ameliorating it.

NO_x is generated from combustion systems in three ways. The mechanisms are referred to as *thermal* (Zeldovich), *fuel bound*, and *prompt* (Fenimore).

15.4 Primary Mechanisms Used

NO_x may be reduced at the source (combustion modification) or after the fact (post-combustion treatment). Combustion modifications comprise thermal strategies, staging strategies, and dilution strategies. Post-combustion methods comprise flue-gas treatment techniques described in Sections 15.5.2 and 15.6.

15.5 Design Basics

15.5.1 Different Forms of NO_x

Nitric oxide (NO) is the most predominant form of NO_x. Most boilers and process heaters generate more than 90% of NO_x as NO. However, gas turbines and other combustion systems that operate with lots of extra air can generate significant quantities of visible nitrogen dioxide (NO_2). NO_2 is reddish-brown in color and responsible for the brown haze called smog. NO, although odorless, oxidizes slowly to NO_2 in the atmosphere. Hence, most NO_x requirements are given as *NO_2 equivalents*.

Hydrocarbons and NO_x react to ground-level ozone. Ozone at high altitude is good because it filters out harmful ultraviolet rays. Ozone at ground level is bad because it interferes with respiration, especially for sensitive individuals such as asthmatics and the elderly. The complicated chemistry among ozone, NO_x, and hydrocarbons is why hydrocarbons and NO_x are strictly regulated. Carbon monoxide (CO) can also participate in the chemistry and is also a regulated pollutant.

15.5.2 NO$_x$ Measurement Units

NO$_x$ is measured in a variety of differing units depending on the source. For example, NO$_x$ from most boilers is regulated as volume concentrations at a reference oxygen condition, for example, 100 parts per million (ppm), dry volume, corrected (ppmdc) to 3% O$_2$. Most NO$_x$ meters analyze their samples after water is condensed. Failure to condense the water before measurement in a dry analyzer could damage the analyzer. Such analyzers are known as extractive analyzers because they must first extract a sample from the stack, condense the water, and then send the dry conditioned sample to the analyzer. *In situ* analyzers read NO$_x$ directly in the hot wet stream. Figure 15.1 shows an analyzer designed to measure the NO$_x$ content *in situ* and report the result in meaningful NO$_x$ units. It uses a nondispersive infrared beam and optical measurement techniques.

The most popular type of post-combustion treatment is selective catalytic reduction (SCR). Ammonia or urea is injected in the flue gas near a catalyst. The net reaction is as follows:

$$2NO + 0.5O_2 + 2NH_3 \rightarrow 2N_2 + 3H_2O$$

Catalysts perform best within a narrow operating temperature range. In some cases, flue-gas tempering or conditioning is required. This may include

FIGURE 15.1
NO$_x$ analyzer (Air Instruments and Measurements, Inc.).

evaporative coolers, air tempering systems, heat exchangers, and so on. Catalyst activity may be adversely affected due to abrasion with ash, high sulfur in the flue gas, or metal poisons.

NO$_x$ is formed in combustion systems in three primary ways. The following provides an overview of each type.

15.5.3 Thermal NO$_x$

The thermal NO$_x$ mechanism comprises the high-temperature fusion of nitrogen and oxygen. This reaction occurs when air is heated to high temperatures such as those that exist in a flame. The reaction is not very efficient. Air contains 79% nitrogen (N$_2$) and 21% oxygen (O$_2$) by volume. Despite this, only 100 ppm or so of NO$_x$ is produced by the thermal NO$_x$ mechanism. Notwithstanding, NO$_x$ is currently regulated to less than 40 ppm in many localities and less than 10 ppm in some regions. Southern California and the Houston–Galveston area are two of the most highly restricted regions for allowable NO$_x$ emissions.

The overall reaction for thermal NO$_x$ formation is as follows:

$$N_2 + O_2 \rightarrow 2NO \tag{15.1}$$

However, the actual elemental mechanism is much more complicated. Nitrogen is a diatomic molecule held together with a triple covalent bond (N°N). This bond takes a lot of energy to rupture, which accounts for the poor efficiency of the overall reaction. Oxygen, however, is a diatomic molecule held together by a double covalent bond (O = O). This bond is much easier to rupture. In fact, oxygen is the second most reactive gas in the periodic table (exceeded only by fluorine, which has a single covalent bond, F-F). These facts make combustion possible but also allow for some attendant NO$_x$ formation. At high temperature, diatomic oxygen forms atomic oxygen.

$$O_2 \rightarrow 2O \tag{15.2}$$

Atomic oxygen is very reactive. The fuel consumes virtually all the reacting oxygen in a combustion system. However, some free radical oxygen collides with diatomic nitrogen in the combustion air to produce NO.

$$O + N_2 = NO + N \tag{15.3}$$

We use the equals sign (=) to indicate that the reaction proceeds on a molecular level, as opposed to the arrow (→), which indicates a net reaction that is a combined series of elemental steps. The atomic nitrogen is also extremely reactive and can attack diatomic oxygen to produce another molecule of NO.

$$N + O_2 = NO + O \tag{15.4}$$

The leftover atomic oxygen goes on to propagate the chain reaction via (15.3). Adding (15.3) and (15.4), we obtain the net reaction given by (15.1).

$$O + N_2 = NO + N \tag{15.3}$$

$$N + O_2 = NO + O \tag{15.4}$$

$$N_2 + O_2 \rightarrow 2NO \tag{15.1}$$

From this chemistry, we can write a rate law. If we presume that reaction (15.3) is the rate-limiting reaction and that oxygen is in partial equilibrium with its atomic form ($\frac{1}{2}O_2 \rightarrow O$), then the rate law becomes:

$$[NO] = \int Ae^{-\frac{b}{T}} \sqrt{[O_2]} [N_2] \, dt \tag{15.5}$$

where the quantities in brackets are the volume concentrations of the enclosed species, A and b are constants, T is the absolute temperature, and t is time. Reaction (15.5) cannot be integrated over the tortured path of an industrial burner because the actual time–temperature–concentration path is unknown. However, the equation does tell us something useful about thermal NO$_x$ formation. Namely, *NO$_x$ is exponentially related to temperature. A small temperature difference makes a big NO$_x$ difference.* This means that hot spots in the flame can dominate NO$_x$ formation. Second, *NO$_x$ is proportional to at least the square root of oxygen concentration.* The nitrogen concentration is less important because it does not change much with little or lots of air. However, the oxygen concentration changes markedly with an increase in combustion air, as it is being consumed in the fuel–air reaction. Finally, *the time at these conditions affects NO$_x$.* Therefore, the highest NO$_x$ will be formed by persistent hot spots in the flame and at high oxygen concentration.

For these reasons, a low-NO$_x$ burner is designed to operate at a temperature that reduces NO$_x$ formation, has a uniform temperature and oxygen pattern within that range, and has a residence time that is conducive to NO$_x$ control.

Special burners have been developed for the purpose of extracting the maximum heat from the fuel while emitting the lowest NO$_x$. Figure 15.2 shows a modern low-NO$_x$ combustor and its principal components.

15.5.4 Fuel-Bound NO$_x$

When nitrogen is bound in the fuel molecule itself, the fuel-bound mechanism operates. The nitrogen must be part of the chemical structure of the fuel. For example, natural gas containing a small percentage of nitrogen gas in the fuel does not produce NO$_x$ via the fuel-bound route because the nitrogen is not bound as part of the fuel molecule. Coal and certain fuel oils

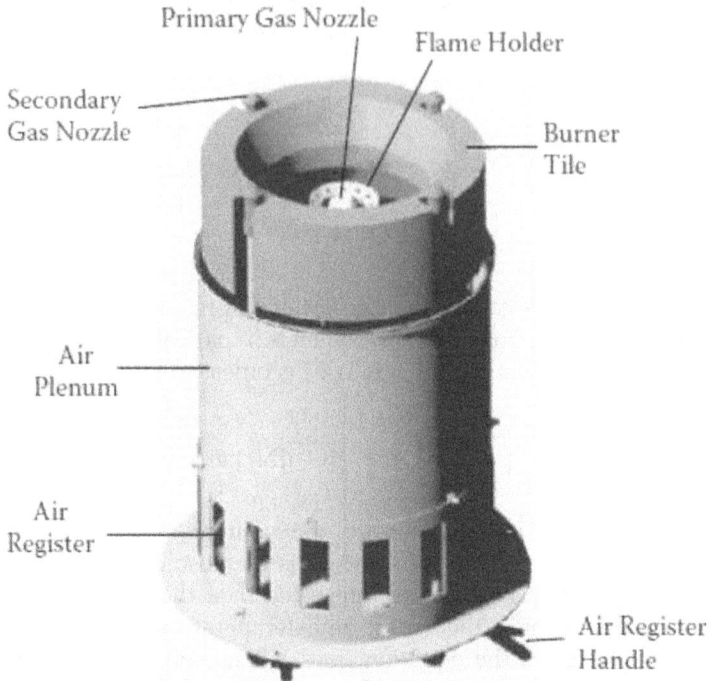

FIGURE 15.2
Low-NO_x burner and components (John Zink Co.).

have nitrogen as part of the fuel molecule, and in those cases the fuel-bound NO_x mechanism may be the predominant NO_x production mode.

As an illustration, consider a hydrocarbon like heavy fuel oil having a small percentage of nitrogen bound in its structure (C_xH_yN), where the subscripts x and y indicate the number of carbon and hydrogen atoms in the molecule, respectively. As the fuel is heated and before it can even react with oxygen, it falls apart to generate some cyano intermediates (HCN, CN). The destruction of a fuel in the presence of heat but not oxygen is referred to as pyrolysis.

$$C_xH_yN \rightarrow HCN, CN \tag{15.6}$$

The pyrolysis reaction is a low-temperature reaction. However, the intermediate cyano species may then react with oxygen to form NO and other species.

$$HCN, CN + O_2 \rightarrow NO + \ldots \tag{15.7}$$

The greater the weight percentage of fuel-bound nitrogen in the fuel, the greater the amount of associated NO_x. However, there is a law of diminishing returns, and at higher nitrogen concentrations things are not as bad as they could be; not all the fuel-bound nitrogen will be converted to NO$_x$. However, for small concentrations of fuel-bound nitrogen, for example, a few hundred ppm in the fuel, the conversion to NO$_x$ is quantitative. Because the pyrolysis reaction is a low-temperature reaction, the peak flame temperature plays a small role in fuel-bound NO$_x$. The more important consideration is access of oxygen to the HCN and CN. Therefore, to reduce fuel-bound NO$_x$, dilution strategies like flue-gas recirculation (FGR), staged air, and fuel dilution are superior to reducing peak flame temperature.

The use of a reference oxygen condition is required for all volume-based measurements. Otherwise, one could simply dilute the effluent stream with air and measure reduced concentrations while making no real reduction in emissions. The factor for dilution correction differs slightly from region to region but is generally of the following form:

$$\text{Corrected NO}_x = \frac{\text{Measured NO}_x \times (20.9 - \text{oxygen reference})}{(20.9 - \text{measured oxygen})} \tag{15.8}$$

For example, 100 ppm NO$_x$ measured at 5.3% O$_2$ works out to be about 115 ppm corrected to 3% O$_2$, for example, $100 \times (20.9 - 3)/(20.9 - 5.3) = 114.7$.

An alternative unit for NO$_x$ from boilers is pounds per million British thermal units (BTU), expressed as lb NO$_2$/MMBTU. With this unit we have several options to consider. Primarily, is the heat release the higher heating or lower heating value? The higher heating value considers the heat from the fuel presuming that the stack gas is cool enough to condense water vapor. For most boilers, the stack is not so cool, but the calculation is usually done on a higher heating value basis anyway.

The lower heating value is often used for process heaters. The lower heating value calculates fuel energy presuming that the stack gas does not condense. Since the lower heating value does not benefit from the heat of condensation, it is lesser by this amount than the higher heating value. For most hydrocarbons, the lower heating value is about 10% lower than the higher heating value. However, one should calculate the difference precisely. For CO (whose combustion generates no water), higher and lower heating values are identical. For hydrogen (whose combustion generates only water), there is a large difference between higher and lower heating value.

For natural gas combustion, presuming a higher heating value basis, 40 ppm at 3% O$_2$ = 0.05 lb/MMBTU, and the relationship is linear. That is 0.10 lb/MMBTU = 80 ppm, ceteris paribus. Process heaters generally use a lower heating value basis, which means that the lb/MMBTU equivalent will be a larger number because we are dividing by a lesser heating value.

Gas turbines are generally regulated to a 15% oxygen reference, while reciprocating engines are regulated on a gram-NO$_2$ per brake-horsepower basis (g/bhp). Some utility boilers are regulated on the absorbed duty (i.e., the heat release less the heat lost out the stack). For these reasons, one must have knowledge of the customary units of the governing regulatory body.

15.6 Thermal-NO$_x$ Control Strategies

Thermal strategies are those that act to lower the peak flame temperature and thus reduce NO$_x$ from the thermal mechanism. One such thermal strategy is FGR. By recirculating a portion of the flue gas into the combustion air, the flame is cooled. A secondary effect of FGR is to reduce the oxygen concentration, again lowering NO$_x$ from the thermal mechanism. The increased mass flow from FGR also adds turbulence and homogenizes the flame, reducing hot spots. The disadvantage of FGR is that fan power is required to recirculate the flue gas. However, FGR can cut NO$_x$ in half. A typical natural gas flame with FGR produces 50 ppm NO$_x$, while the flame without FGR produces about 100 ppm. Generally, no more than about 25% FGR can be recirculated in a conventional burner before stability problems occur.

Steam or water can be added to the flame by means of an injection nozzle. The nozzle is moved to a location that does not interfere with combustion but cools off the flame. This strategy costs little in capital cost to implement. However, the water or steam carries heat away from the flame that is not recovered, so thermal efficiency losses result.

15.7 Dilution Strategies

FGR acts primarily to cool the flame and secondarily as a dilution strategy for the oxygen in the combustion air. Recirculating flue gas to the fuel side for gas fuels can be more effective than FGR in reducing NO$_x$ for several reasons. First, gaseous fuels are usually supplied at pressures of 40 psig or above for industrial settings. This fuel energy may be used in an eductor arrangement to pull flue gas from the stack. When such a strategy is feasible, fuel dilution requires no external power. Second, diluting the fuel directly reduces concentrations of HCN and CN that occur on the fuel side, thus reducing fuel-bound and prompt NO$_x$. Diluting the fuel or air stream with any inert agent, be it nitrogen, CO_2, noncombustible waste stream, or steam, reduces NO$_x$ from thermal and dilution mechanisms. Care must be taken not to reduce the fuel or oxygen near or below their flammability limits—otherwise the flame will become unstable or go out. In extreme cases, burner instability can result in an explosion if a flammable mixture fills the furnace and suddenly finds a source of ignition.

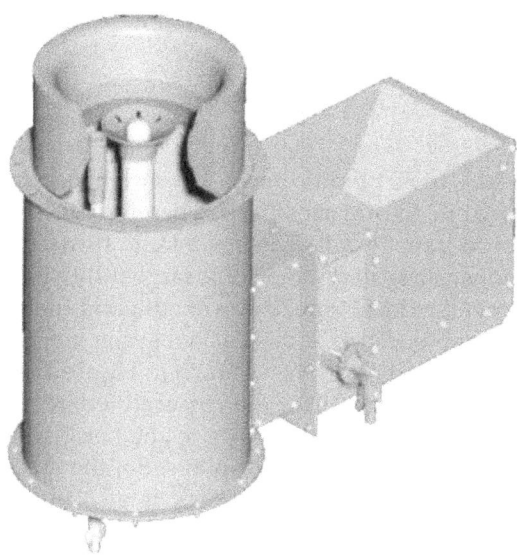

FIGURE 15.3
Staged combustion burner (John Zink Co.).

15.8 Staging Strategies

Rather than mixing all the fuel and air together at once in a hot combustion zone, either the fuel, air, or both may be staged along the length of the burner. The stepwise addition of fuel (two or three stages are sufficient) delays mixing and allows for some heat transfer to the surroundings before further combustion takes place. Air staging is generally considered more effective to reduce fuel-bound nitrogen, while fuel staging is more effective at reducing thermal NO$_x$. Figure 15.3 shows a staged combustion burner designed specifically for NO$_x$ reduction.

15.9 Post-combustion Strategies

Selective noncatalytic reduction (SNCR) uses ammonia (or an ammoniacal agent) to reduce NO$_x$. At some temperature between 1400 and 1800°F, ammonia dissociates to form NH$_2$.

$$NH_3 = NH_2 + H \tag{15.9}$$

NH_2 is a short-lived and very reactive species that reduces NO to nitrogen and water.

$$NH_2 + NO = N_2 + H_2O \qquad (15.10)$$

SNCR can reduce NO_x to 50 ppm or lower. However, such reaction temperatures are found within the furnace itself. Therefore, to provide adequate mixing and residence time, SNCR requires a large furnace (e.g., coal-fired, and municipal solid waste systems and some large utility boilers). Most SCR catalysts are base metal oxides, especially vanadia and titania deposited on an alumina honeycomb surface. A typical honeycomb-type catalyst block containing exotic base metal catalysts is shown in Figure 15.4.

By adding a catalyst, one can lower the required temperature window to 500–750°F. These temperatures occur close to the stack in process heaters and within the air-preheaters of larger boilers. So, the size of the furnace is not such an important factor. The strategy is also more effective than SNCR, generating 90% NO_x reductions or greater. The important steps are adsorption of ammonia and NO_2 onto the catalyst surface (X–Y). NO_2 may be formed rapidly from NO by oxygen on the catalyst surface or in

FIGURE 15.4
Post-combustion honeycomb catalyst (Bremco).

the gas phase. Water on the surface protonates the ammonia to NH$_4$. The essential chemistry is:

$$NH_3 + -X \rightarrow (\text{with moisture}) X - NH_4^+ \qquad (15.11)$$

$$NO_2 + -Y \rightarrow X - NO_2 \qquad (15.12)$$

The adjacent sites hold the ammonia and NO$_2$ in proximity, where they quickly react, restoring the catalyst surface for additional reactions. An electron from the surface is required to balance the reaction.

$$X - NH_4^+ + Y - NO_2 + e^- = X - Y + N_2 + 2H_2O \qquad (15.13)$$

15.10 Operating/Applications Suggestions

An intelligently designed NO$_x$ control system starts with the accurate determination (or estimation) of the NO and NO$_2$ that is or will be produced from the source.

Accurate sizing and specification of low-NO$_x$ burners requires consideration of fuel properties, furnace operating temperatures, excess oxygen conditions, and knowledge of the service application. This almost always requires detailed conversations between the burner vendor and the end user.

Likewise, SCR systems require detailed conversations between the end user and the SCR system supplier. The catalyst can be rendered ineffective by physical blinding with inert particulate, abrasion, or poisoning by certain heavy metals or sulfur. An inventory of any possible fouling or poisoning agents must be derived first by analyzing the fuel, its metals content, and its propensity to form oxides or produce partially burned or unburned carbonaceous compounds and comparing the result to known fouling agents for the proposed catalyst. Possible remedies include, among others, removal of fouling agents before the catalytic stage, use of a sacrificial precatalyst, or more frequent catalyst replacement.

In SCR or SNCR systems, unreacted ammonia that slips through the system is termed *ammonia slip*. Ammonia slip is more easily controlled on baseloaded (steady-state) operations. In such a case, the ammonia injection rate can be determined by experience and testing, then maintained in an optimum range. Feedback controls can sometimes be used to adjust the ammonia rate, however, to date, these have proven to be slow to respond. Usually, some ammonia slip is tolerated, and larger NO$_x$ reductions are possible if higher ammonia slip rates are acceptable. Some regulatory districts are putting limitations on the total allowable slip, thus complicating NO$_x$ control.

16

Thermal Oxidizers

Dan Banks

Banks Engineering, Inc., Tulsa, Oklahoma

16.1 Device Type

Thermal oxidizers (TOXs) are used to destroy objectionable hydrocarbons contained in waste streams from manufacturing plants. The wastes may be solids, liquids, or vapors. They are usually generated continuously—otherwise landfill may be economically preferred for solids and liquids, while emergency flares might be preferred for destruction of many waste gases. TOXs are designed to use heat energy to convert hydrocarbon contaminants to carbon dioxide and water vapor, and contaminant metals to their oxide form, under controlled conditions.

16.2 Typical Applications

TOXs are used to control combustible contaminant emissions from dozens of sources. Major areas include printing operations; chemical, and hydrocarbon processing; painting, coating, and converting; distillation; sludge drying; soil remediation; plasticizer emissions control; extruder emissions; and textile manufacturing.

They are often used after wet scrubbers where the gas stream contains both water-soluble and hydrocarbon emissions. They are often followed by wet scrubbers where the volatile organic compound (VOC) is halogenated and, upon combustion, can form inorganic acids such as hydrochloric acid (HCl).

In general, if the source emits a combustible VOC that is not economical to recover, it is a candidate for control by a TOX.

16.3 Operating Principles

A TOX simply heats the waste material in the presence of air to allow the hydrocarbon molecules present to burn (oxidize at elevated temperature). The simplest TOX consists of a burner, a holding chamber (furnace), and a stack (to duct the combustion products to atmosphere). Furnace temperature can range from 500 to 2500°F, depending on TOX design and the degree of hydrocarbon destruction needed. If 99% of the incoming hydrocarbons are destroyed, the TOX efficiency is 99% (expressed as 99% destruction and removal efficiency [DRE]). Usually, natural gas or other auxiliary fuel is ignited in the burner to heat up the TOX and often to supplement the heating value of the waste stream(s) to ensure proper temperature control. If the waste is rich in hydrocarbons, extra air or sometimes water sprays are used to prevent overheating. Various methods have been developed to reduce fuel usage, keep generation of NO_x and other pollutants low, recover available heat from the combustion products, and remove any particulate or acid gas (HCl, SO_2) formed during waste destruction.

To make the best use of this application of heat energy, the TOX is usually lined with insulating refractory material.

Figure 16.1 shows a TOX used for the control of noncondensable gases from a paper pulp mill. The unit consists of a specially designed burner, burner controls, insulated combustion chamber, and temperature controls.

16.4 Primary Mechanisms Used

Reacting hydrocarbons with oxygen results in release of energy. An example is the oxidation of natural gas (methane):

$$CH_4 + 2O_2 \rightarrow CO_2 + 2H_2O$$

where one molecule of methane combined with two molecules of oxygen forms two molecules of carbon dioxide and two molecules of water vapor.

If air (79% nitrogen) is used to provide the oxygen, other gases go along for the ride:

$$CH_4 + 2O_2 + 7.5N_2 \rightarrow CO_2 + 2H_2O + 7.5N_2$$

This is the balanced stoichiometric equation for combustion of methane with air and is typical of the combustion equation that is used in designing a TOX system for destruction of any hydrocarbon. When one pound of methane is burned in a TOX furnace, the product gases exit at much higher temperature—the net heat released by burning this one pound of

FIGURE 16.1
Noncondensable gas thermal oxidizer. (Banks Engineering, Inc.)

hydrocarbon is 21,280 British thermal units. The methane reaction written above would produce products at over 3000°F, requiring special furnace construction, so extra air, or water sprays (or a low heating-value waste stream) would be added to produce products at 2000°F or lower.

High-temperature oxidation proceeds at a higher rate at higher temperatures, but as less and less of the subject hydrocarbon is left, the destruction rate slows. Operating the TOX furnace at a higher temperature increases the DRE in each furnace or allows use of a smaller furnace to achieve the original DRE. Some hydrocarbons are easy to destroy, requiring low temperatures and little retention time in the furnace (small furnace). Others require higher temperatures and longer reaction times for the same DRE. For instance, 99.99% DRE of hydrogen sulfide requires about 1300°F and 0.6 s retention time, while 99.99% DRE of dichloromethane requires about 1600°F and 2 s retention time.

If a chlorinated hydrocarbon is oxidized, the raw unbalanced equation might look like this:

$$CH_2Cl_2 + O_2 + N_2 \rightarrow CO_2 + H_2O + HCl + N_2$$

In this case, dichloromethane burns to produce carbon dioxide, water vapor, and HCl. If enough HCl is formed, discharge directly to atmosphere would not be permitted and the combustion products would be cooled and reacted with a chemical such as caustic (NaOH) to remove most of the HCl. The same is true when the waste contains hydrogen sulfide (H_2S forms sulfur dioxide [SO_2]). If the waste contains ash or dissolved solids (like salt), then the combustion products will contain particulate matter. Excessive particulate matter must be removed (Venturi scrubber, electrostatic precipitator, bag filter, etc.) before the combustion products are discharged to atmosphere.

16.5 Design Basics

A TOX always includes these items:

1. Auxiliary fuel burner
2. Air source (blower or natural convection)
3. Furnace (temperature-controlled chamber where the oxidation reactions occur)
4. Stack (to direct the combustion products to atmosphere)
5. Control system (to verify proper operation and control excursions)

Figure 16.2 shows a typical TOX used to destroy hydrocarbon emissions. The system consists of the burner (to the left, top), the lined combustion chamber, a down-fired quencher, and exhaust ductwork (lower right).

FIGURE 16.2
Thermal oxidizer for VOC control. (Banks Engineering, Inc.)

Depending on the waste(s) to be treated, a TOX system can also contain the following:

1. Waste heat boiler (cools the combustion products, recovering the heat generated in the furnace by evaporating water to make steam for other uses)

2. Wet scrubber (packed bed, Venturi, or spray scrubber, where acid gases and/or particles are removed from the combustion products)

3. Dry scrubber (bag filter, electrostatic precipitator, etc., where particulate matter and sometimes acid gases are removed from the products)

4. NO_x (nitrogen oxides) reduction hardware (catalytic, noncatalytic, or wet scrubber NO_x removal processes)

5. Preheat exchanger (usually shell/tube or plate/plate device with combustion products on one side and waste or combustion air on the other side, where heat is recovered and directed back into the TOX furnace to save auxiliary fuel)

6. Catalyst bed (speeds oxidation of particle-free waste gas, allowing lower operating temperature for the same DRE as a noncatalytic TOX)

7. Concentration methods to eliminate some of the inert gages in a waste stream before sending the residual hydrocarbons to the TOX (often accomplished with a heat regenerated zeolite bed)

In Figure 16.3, we can see a TOX system arranged as a compact unit complete with a local control panel.

Direct thermal units burn fuel gas or fuel oil to ensure waste ignition and maintain desired furnace temperature, when necessary. A recuperative TOX system adds a heat exchanger to transfer heat from the combustion products to the incoming waste gas or combustion air, reducing fuel consumption. Direct thermal TOX systems can be used to handle waste liquids and waste gases.

A variation on the direct flame type TOX is a design provided by Alzeta (California). Figure 16.4 shows a facility using an Alzeta 500 cfm flameless TOX equipped with an alloy C-276 quencher and fiberglass packed tower. These flameless designs incorporate special internal porous modules that provide a combustion surface instead of a flame front as in a conventional burner. Such designs in theory provide for superior combustion control and fuel–air mixing, resulting in decreased emissions and higher thermal efficiency.

Catalytic TOX units may also fire fuel oil or fuel gas, but smaller ones may use electrical resistance heating instead. The catalyst reduces the temperature needed for a specific DRE, reducing fuel consumption. These units usually include a heat exchanger to further reduce fuel demand by transferring

FIGURE 16.3
Thermal oxidizer. (Alzeta Corp.)

FIGURE 16.4
Flameless thermal oxidizer. (Alzeta Corp.)

heat from the combustion products to the waste gas before furnace entry. Catalytic TOX units can be used to handle particulate-free waste gases containing small concentrations of hydrocarbons; excessive temperature and entrained dust interfere with the catalyst.

A catalytic TOX is shown in Figure 16.5. This one controls VOC emissions from a semiconductor manufacturing facility.

Regenerative thermal oxidizers (RTOs) route the waste gas through thermal-mass-packed beds for heat recovery, allowing exceptionally low fuel requirements, even for lightly contaminated waste air streams. RTOs are commonly used to treat large flows of air containing traces of hydrocarbons. Many operate with little or no auxiliary fuel due to the excellent heat recovery offered by packed beds, but waste containing too much hydrocarbon can overheat RTOs.

Figure 16.6 is a diagram of the functional components of a typical RTO. A series of dampers alternately feeds gases to the appropriate chamber

FIGURE 16.5
Catalytic thermal oxidizer. (Alzeta Corp.)

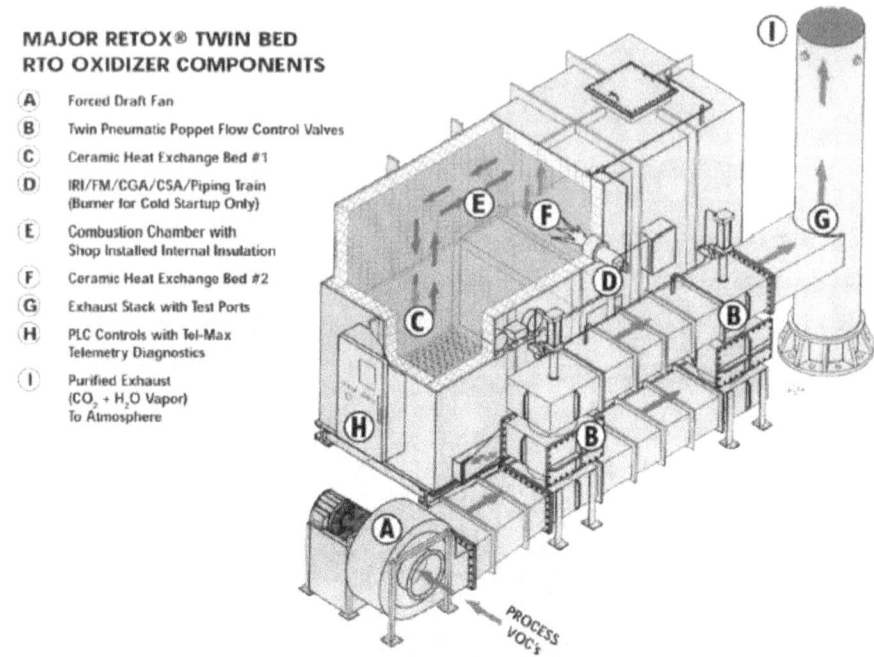

**MAJOR RETOX® TWIN BED
RTO OXIDIZER COMPONENTS**

(A) Forced Draft Fan

(B) Twin Pneumatic Poppet Flow Control Valves

(C) Ceramic Heat Exchange Bed #1

(D) IRI/FM/CGA/CSA/Piping Train
     (Burner for Cold Startup Only)

(E) Combustion Chamber with
     Shop Installed Internal Insulation

(F) Ceramic Heat Exchange Bed #2

(G) Exhaust Stack with Test Ports

(H) PLC Controls with Tel-Max
     Telemetry Diagnostics

(I) Purified Exhaust
     (CO₂ + H₂O Vapor)
     To Atmosphere

PROCESS VOC's

FIGURE 16.6
Regenerative thermal oxidizer. (Adwest Technologies, Inc.)

for either preheating or combustion. The thermal mass of the refractory or ceramic fill retains the heat sufficiently to allow reasonable time between cycles. The VOCs enter at A and are directed through a plenum containing dampers, B, which permit switching gas flows between the chambers, C, which contain thermal mass. A supplemental burner, D, provides any additional heat to sustain combustion (if required). The hot gases exit through the alternative thermal mass, F, thereby heating it. The combustion products leave through the stack, G. The control panel, H, switches between chambers so that the desired combustion conditions are maintained.

An actual installation may look something like the installation shown in Figure 16.7.

RTOs are also used to control the VOC emissions from dryers that dry material that may, during heating, evolve VOCs such as with the wood chip dryer shown in Figure 16.8. Typically, primary particulate separation is accomplished at the dryer outlet using dry cyclone collectors and perhaps the addition of a wet electrostatic precipitator (WESP). Once the particulate is removed, the VOCs are controlled in the RTO. The discharge stacks with vortex shedding strakes are shown at the far left of the picture.

FIGURE 16.7
RTO type oxidizer. (Adwest Technologies, Inc.)

16.6 Operating Suggestions

Claus sulfur recovery plants generate a waste gas containing H_2S, CO, water vapor, and inert gases. Waste flow is steady. TOX operation ranges from 1200 to 1500°F with a furnace retention time of 0.6–1.0 s. A vertical refractory-lined furnace is often used, allowing a shorter stack to improve dispersion of the combustion products (which contain SO_2). The furnace/stack generates draft, and burner operation does not need a combustion air blower. A waste heat recovery boiler may be added, in which case the furnace is horizontal, and a combustion air blower is added.

Pharmaceutical plants generate air-rich or nitrogen-rich waste gases from various batch reactors. Waste flow and composition may change suddenly, so burner control requires special care. A pharmaceutical TOX may operate at 1600–2000°F with 1-s retention time, depending on the waste components and performance required. A waste heat boiler and wet scrubber (for HCl produced by combustion of chlorinated compounds) are often used.

FIGURE 16.8
RTO on chip dryer. (A.H. Lundberg)

Kraft pulp mills generate several acidic waste gases during papermaking. Several of the waste streams can contain both oxygen and hydrocarbons, presenting flashback problems. Stainless burner and waste duct construction is common. A wet scrubber is used to remove SO_2, which is generated during combustion of the H_2S and similar compounds in the waste gas. A turpentine byproduct may be burned in special guns to reduce firing of natural gas or fuel oil.

A TOX system must be designed to handle the full range of waste types, waste flows, and waste compositions. If errors are made, the system may run short of fuel, air, reaction volume, scrubbing capacity, or other critical items. Poor waste destruction can result, but damage to the TOX unit or even upstream process equipment is certainly possible.

The minimum operating controls needed are aimed at preventing thermal damage or explosions. A common design standard is provided by the National Fire Protection Association. With wastes that vary in flow or heating value, additional controls may be required for quick adjustment of fuel or air to always maintain on-spec operation.

High-temperature operation requires special attention to the various refractories, stainless steels, paints, and plastic used for construction, since an error in this area can quickly lead to catastrophic failure. Temperature control is always important, especially where catalyst is used to improve waste destruction, since excessive temperature can destroy catalyst quickly.

Refractories can be damaged by abrupt temperature changes. Slow start-ups (200°F temperature rise per hour) are typical. Ceramic fiber blanket refractory linings may be heated much more quickly. Periodic refractory inspection (usually once per year) is suggested, to allow repair of damage areas before they expand to create serious problems.

Some wastes form SO_2, HCl, or other acidic compounds when burned. These are normally harmless when hot, but areas where the combustion products can cool to 200–300°F may be subject to severe corrosion if the acid–gas dewpoint is reached. In units with acidic combustion products, the TOX furnace should be protected with weather shielding or be located indoors. Wet scrubbers are often applied to TOX units to control the acid gases produced. If particulate is also present or is created through the combustion process, particulate control devices such as Venturi scrubbers, dry scrubbers, or WESPs are often used.

The presence of suspended ash, dissolved salts, or other particulate-producing compounds may require special design to avoid blinding of waste heat recovery surfaces and damage to refractory or excessive emissions.

17

Tray Scrubbers

17.1 Device Type

Tray scrubbers are wet scrubbers that use a tray or multiple trays containing openings through which gas is accelerated to subsequently mix with scrubbing liquid, thereby enhancing particulate removal and gas absorption.

17.2 Typical Applications and Uses

Tray scrubbers are used for the control of particulate greater than approximately 10 μm at loadings less than 1–2 grs/dscf (grains per dry standard cubic foot) and to absorb soluble gases. They are also often used for cooling gas streams and to subcool gases.

In years before the Clean Air Act (1970), tray scrubbers were used to control the emissions from lime kilns, lime sludge kilns, boilers, tanker inert gas systems, sludge incinerators, and similar applications where particulate and soluble gases (usually acidic gases) must be removed simultaneously. As the air pollution control regulations tightened, higher efficiency wet devices, such as Venturi scrubbers, and dry devices, such as baghouses, became more popular. Tray scrubbers continue to be used as gas cleaning and conditioning devices, often in concert with other devices.

Tray scrubbers are not used where large-diameter particulate may plug the openings in the trays. They are also avoided where the loading of dust is high (above 2 grs/dscf) for the same reason. You often see tray scrubbers successfully used after primary particulate devices. For example, they are used for gas cooling and plume suppression after Venturi scrubbers on municipal sludge incinerators and after baghouses to remove SO_2 and HCl from waste incinerators.

17.3 Operating Principles

Tray scrubbers use one or more punched, perforated, drilled, or woven trays (usually flat) over which scrubbing liquid is passed. Gases containing particulate and absorbable gases are passed through these openings wherein the gas is accelerated, increasing its kinetic energy. The gas impacts into this liquid passing over the tray and in doing so transfers its energy into the liquid, causing a froth, or bubbling, turbulent zone of high surface area. The liquid typically passes over the tray, usually over a weir to stabilize the liquid depth, then descends through a downcomer to the next lowest tray or leaves the scrubber vessel via an external drain.

Figure 17.1 shows a tray scrubber in isometric view. It consists of a low-level gas inlet, a spray assembly to clean the face of the tray, a removable flanged grid assembly, a liquid inlet and distribution weir box, internal downcomer, a vane type droplet eliminator (the turbine-shaped device at the top), and the gas outlet at the top.

Some trays are equipped with baffles immediately opposite the tray holes or perforations. These are called *impingement tray scrubbers* because the gas impinges on the baffles. These trays generally offer greater particulate removal given the impaction action. Figure 17.2 shows the basic components of a tray scrubber. The gas inlet injects the gas stream into a zone below the tray section. Quite often preconditioning sprays (as shown) are used to saturate the gas stream so that the gas stream characteristics are more uniform as the gas enters the tray openings. A uniform gas flow serves to provide a more stable pressure drop across the width of the tray. These sprays may be oriented downward, upward, or both. An upward spray serves to clean the lower side of the tray. The trays are oriented horizontally and must be levelled to achieve uniform liquid distribution. The liquid is distributed using weirs and downcomers. The "end weir" serves to hold up liquid on the tray and influences the pressure drop across the tray since the rising gas must penetrate that liquid depth. The downcomer allows the scrubbing liquid to exit the tray and move to the next lower tray (or to the sump as is the case with the lowest tray) while preventing gases from "sneaking" around the tray and as a result reducing the tray contact efficiency.

The downcomer of a customized tray is shown in Figure 17.3. The downcomer baffle serves to prevent the gas flow from bypassing the tray by going up the downcomer. Obviously, the liquid depth in the downcomer seal leg must be deep enough to provide a suitable liquid seal.

Trays with larger openings are called *sieve tray scrubbers* and are used primarily for gas absorption. If the gas velocity is insufficient to keep the liquid on top of the tray, the liquid can drain or weep through the trays. These types are called *weeping sieve tray scrubbers.*

If you read Chapter 14, you probably noticed that some modern flue gas desulfurization systems use a hybrid tray/spray scrubber to reduce the liquid

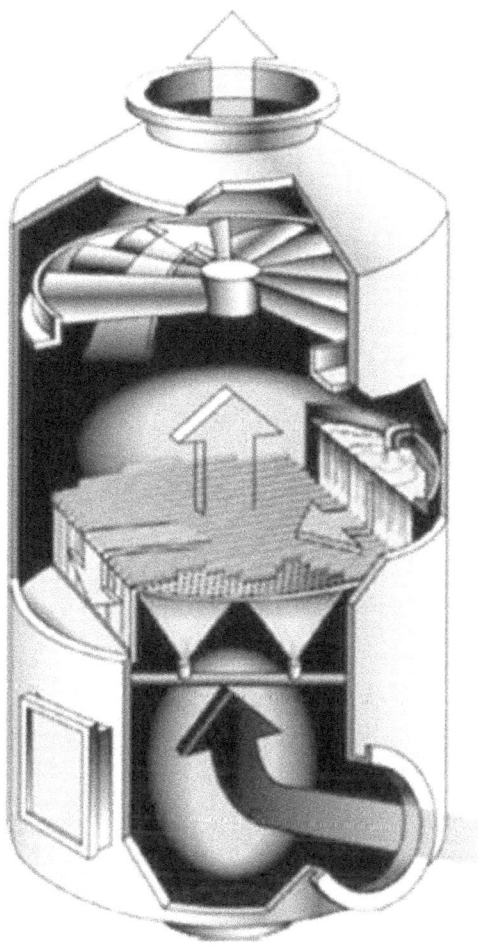

FIGURE 17.1
Impingement tray scrubber (Sly, Inc.).

rate requirements, thereby lowering the pumping horsepower and operating cost. These trays are open, weeping type designs that afford some momentary holdup of scrubbing liquid. This increases the liquid loading per unit volume at the tray and enhances mass transfer in part by reducing the diffusion distance from the gas molecule to the droplet. The stirring of the liquid also improves mass transfer. The liquid gets to the tray, however, using spray nozzles. Eliminate the spray nozzles and increase the gas velocity further, and you get a fluidized bed scrubber. Now introduce a novel way to stabilize this fluidized bed, and you get the patented *RotaBed®* scrubber.

Another variety of the tray scrubber is the bubble cap tray scrubber. These trays have numerous modules called bubble caps that divert the gas into the

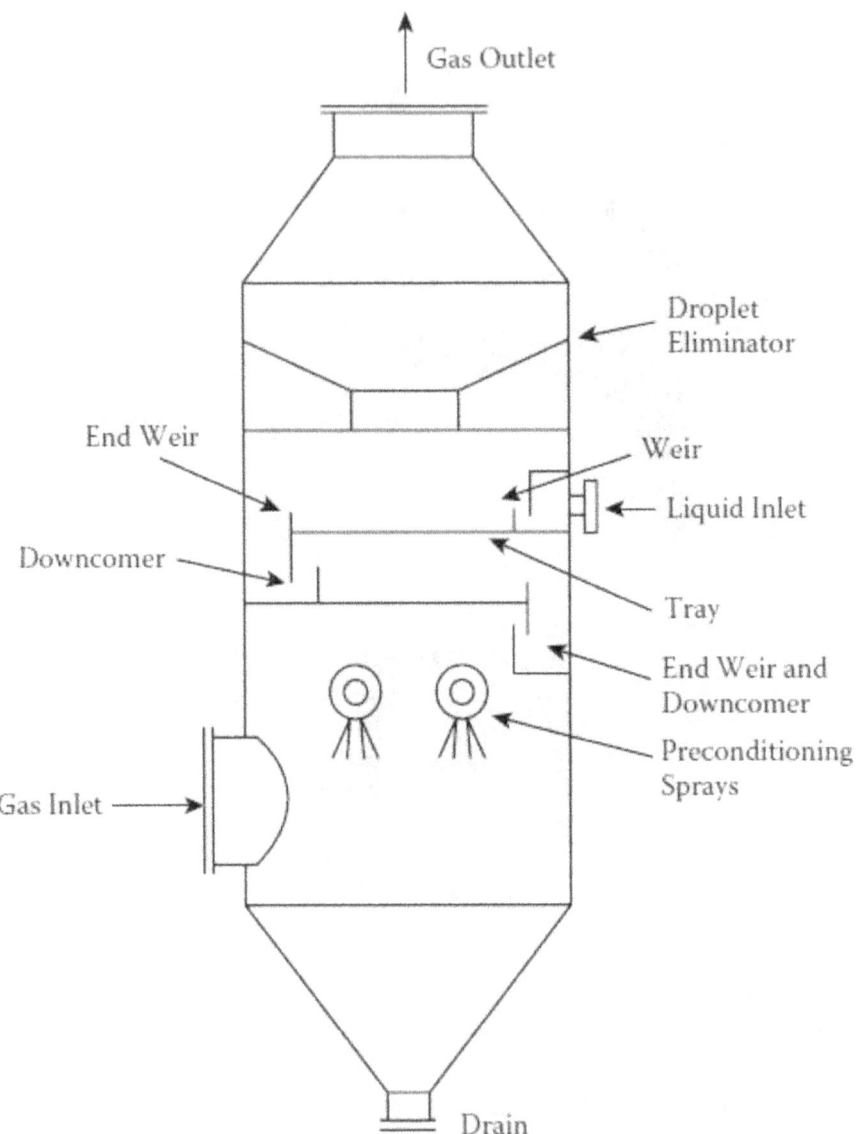

FIGURE 17.2
Tray scrubber basic components.

liquid to create the enhanced surface area. The tray openings are generally larger than in a conventional perforated tray, therefore the bubble cap tray is more plugging resistant. Figure 17.4 depicts a bubble cap tray assembly. These individual caps are often welded to the tray surface to provide secure attachment. Because the hole sizes are greater, the tray strength is somewhat

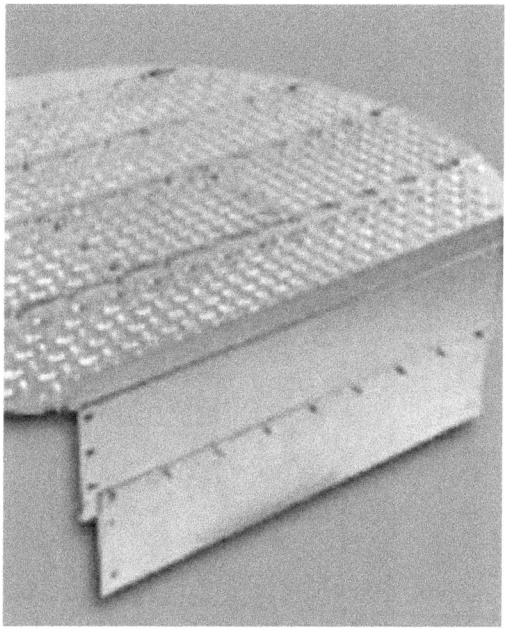

FIGURE 17.3
Tray with end weir and downcomer. (Koch-Glitsch, Inc.)

reduced, therefore these trays are often supported from below. The result is a very sturdy and efficient mass transfer surface.

Some trays are equipped with movable discs that are influenced by the motion of the gas onto their surface. These designs are called *valve tray scrubbers,* and the trays are called valve trays because the discs function as valves.

FIGURE 17.4
Bubble cap tray. (RVT Process Equipment, Inc.)

FIGURE 17.5
Grid type packing tray. (Koch-Glitsch, Inc.)

As the gas flow varies, these discs or valves automatically compensate within their design range. This feature can be of great value if the gas density or flow rate varies significantly.

It can be argued that a cross between a tray and a packed tower is the structured grid scrubber. It uses modules or sections of packing in the form of grids. These grids are stacked in the tower and are irrigated with scrubbing liquid. The structured grid surface extends (enlarges) the liquid surface, thereby improving mass transfer. Figure 17.5 shows a popular type of structured packing. These devices are operated much like a packed tower, yet the contact elements (the structured grid) are installed and replaced in modules.

17.4 Primary Mechanism Used

For absorption of gases, tray-type scrubbers use the high-velocity jets formed as the gas passes through the tray openings to shear the scrubbing liquid into a high surface area dispersion that enhances mass transfer.

For particulate control, the primary mechanism is impaction.

17.5 Design Basics

The perforations in tray scrubbers can range from approximately 1/8 to 1 inch. If the gas flow is sufficient, any of these type scrubbers will flood or operate so that the liquid cannot descend. Designing them, at least in part, revolves around operating the scrubber below this flooding velocity. In contrast, fluidized bed scrubbers operate at higher velocities nearly at flooding speeds.

The typical tray scrubber uses a vessel velocity of approximately 8–10 ft/s vertical velocity (free space). The perforation sizes and spacing can vary. The number of holes will vary the net open area. The following chart shows the approximate open area of some tray designs:

Perforations	Open Area (%)	Dry Pressure Drop
$\frac{1}{4}''$ on $\frac{3}{8}''$ ctrs	40	1.25–1.75″ water column
$\frac{3}{16}''$ on $\frac{3}{8}''$ ctrs	22	1.0–1.5
$\frac{1}{16}''$ on $\frac{3}{16}''$ ctrs	11	1.5
$\frac{1}{2}''$ on $1\frac{3}{32}''$ ctrs	22	1.5–2.0

Inlet weirs are often 3–4 inches high and the liquid enters free-flow (i.e., very low-pressure drop). The weirs are installed so that the weirs are level. This is important to keep a uniform liquid level across the weir since the pressure drop (and resulting performance) is a function of the dry pressure drop of the tray plus the pressure drop associated with the liquid. The liquid usually adds 0.25–1-inch water column pressure drop, per tray, added to the dry grid pressure drop above. The liquid static depth on the tray, however, may be higher than the indicated pressure drop. This occurs because the turbulent mixture above the tray becomes aerated and therefore has a lower density than the static liquid.

Liquid downcomers are usually sized for approximately 80–100 gpm per running foot of weir. This keeps the liquid level uniform across the tray. This brings up an interesting point. If the liquid rate must be higher (given the loading of the contaminant or the cooling duty), the tray scrubber can be configured to have the liquid enter down the center of the tray and have the liquid flow to either side. The latter type of design is called a split flow or dual flow-type tray. Curved or partial circumferential weirs can be used to allow the proper liquid flow rate.

The volumetric flow rate of liquid in the downcomer is usually limited to about 2 ft/s, particularly if an upper tray discharges into a lower tray. High-drain velocities can cause disrupted liquid flow to the tray and a loss in efficiency.

Single trays can be used, but most often the trays are installed in quadrants so that the trays may be removed through manholes. The trays are bolted down or are held in place by wedges and keeper plates.

In each of these designs, the vendors have developed efficiency parameters through experience or testing that equates the performance of the tray as it relates to a theoretical tray (see Chapter 1). Once the number of transfer units required is determined, the number of tray stages can be determined by dividing the relative efficiency per tray into the required number of transfer units required. A typical real tray has an efficiency of approximately 80% of a theoretically perfect tray. Therefore, a typical tray produces about 0.8 transfer units (the actual number varies considerably by tray design).

The vendors of these trays and tray scrubbers can match the tray design to your performance requirements. Although textbook design may reveal how many tray stages you may need, it is best to contact the tray vendors for a proper evaluation.

17.6 Operating Suggestions

Because tray scrubbers use perforated plates or small openings, care should be taken when applying them to high (above 5 grs/dscf) dust-loading applications. In these circumstances, either a prescrubber (such as a Venturi scrubber) is used or the lowest tray is sprayed from below to help keep it clean. Fluidized bed-type scrubbers can often be used instead because they have larger gas ports and are more resistant to plugging. Similarly, tray scrubbers generally operate best at under 1% total suspended solids in the recycle liquid. If you must run at elevated solids, bring that to the attention of the scrubber designer so you do not create a maintenance and operations headache.

Most tray scrubbers use trays configured in removable plate form. These are usually sized so that one or two people can handle them (when the trays are clean). When designing the scrubber installation area, space should be allowed to remove the plates safely. Extra wide access platforms allow space to place the trays after they are removed and can be a handy addition to any installation.

Vessel access doors are frequently too small, requiring the tray panel to be tilted on a diagonal to remove it. If possible, the door width should be at least the width of the tray. If you ask for this, most vendors can accommodate you.

Downcomers and weirs, if used, are a source of plugging and buildup in many applications. Any internal inspection should include a thorough checking of the weirs and downcomers. These areas should be cleaned and repaired as required because they are an integral part of the tray operation. Plugged downcomers can cause improper liquid distribution on the subsequent tray, resulting in a reduction in efficiency. Worn, corroded, or plugged weirs produce improper liquid distribution on the tray, also reducing efficiency.

If upon inspection of a tray one notices an uneven distribution of wear or buildup, it could be caused by an insufficient liquid depth on the tray. Another telltale sign is low (below designer's set point) pressure drop at design flow rates. It is often possible to add an end weir (or increase the height of an existing one) to increase the liquid retention depth. This will also increase the pressure drop on the tray.

The single most important aspect of tray scrubber installations is that the tray is horizontal. Liquid on the tray is inherently trying to stay level. If the tray is not level, the liquid depth can vary across the tray. This can cause a resistance to flow imbalance, producing areas of high velocity and poor contact. When the scrubber is installed and at service periods thereafter, checks should be made to ensure that the trays are level or in accordance with the vendor's specifications.

If you have performance problems with tray scrubbers, various vendors and some consulting firms have analytical equipment to inspect the scrubber and isolate the problem. These devices look for variation in gas and liquid flow patterns. Once these patterns are known, corrective measures can be taken.

18

Vane Type Scrubbers

18.1 Device Type

Vane type scrubbers are wet scrubbers that use one or more stationary vanes through which or within which the contaminant gas streams mix with scrubbing liquid. There are many innovative designs within this category. They are used to remove particulate in the 5 μm and larger size range and provide moderate gas absorption capability. These scrubbers are low- to medium-energy input devices and find themselves in use where the particulate loading is under 4–5 grs/dscf (grains per dry standard cubic foot) and the particle size is 10 μm or above.

There are several interesting and efficient vane type scrubbers currently being provided by vendors worldwide.

18.2 Typical Applications

Vane type scrubbers are often found in use on rotary dryers, grinders, mullers, and similar devices producing relatively large particulate. At higher pressure drops (above 10–15 inches water column), cage type units have been used on nonferrous metals (i e, aluminum, and zinc) remelt furnaces to remove residual metallic dusts, and so forth.

There are hundreds of vane type scrubbers in daily use. Some, in recent years, have been followed by wet electrostatic precipitators or other devices for enhanced capture of submicron-sized particulate.

If the gas stream contains sticky particulate, certain vane designs can rapidly plug. Care must be taken to wet all surfaces fully and constantly, and this can be difficult to accomplish. Given that they require centrifugal action, the gas speed in this type of scrubber is of importance. Turndown ratios of approximately 25% are common, below which point some reduction in efficiency may occur.

18.3 Operating Principle

Vane type scrubbers basically use a multiplicity of Venturi scrubber sections to accelerate the gas stream, causing liquid that is dispersed on the vanes to shear into tiny target droplets. In addition, the designs use centrifugal force to throw the gas stream toward (in most cases) the vessel wall. In some designs, the vanes direct the dispersion of droplets inward into a high droplet density cloud that is configured to maximize impaction and interception.

18.4 Primary Mechanisms Used

Centrifugal and impaction forces are most applied in vane type scrubbers. The vanes may be configured in a near horizontal plane (with vanes oriented much like a gas turbine blade) or as a vertical cage of vanes similar in appearance to a squirrel cage blower impeller. Other designs use various vane combinations but share these primary separation mechanisms.

The vane blades are typically close together, forming a multiplicity of Venturis, and are angled to deflect the gas stream in a way that increases its rotational motion either outward toward the vessel wall or inward into a confined spray zone. These blade groups are often sprayed with the scrubbing solution or the liquid can cascade onto the vane surface. Depending on the orientation of the vane group, the liquid may produce a froth somewhat like a fluidized bed scrubber. A significant difference between the contact zones is that the froth in a vane type scrubber usually proceeds from the vane area and is thrown outward against the vessel wall. In fluidized bed scrubbers, the froth or fluidized zone descends back directly into the gas path and helps create and maintain the froth zone.

Figure 18.1 shows a vane type scrubber wherein the vane group is horizontal. The gas usually enters tangential below this vane group to impart a cyclonic motion and provide some centrifugal separation as in a cyclone collector. The gases then move vertically into the vane group area, where the angle of attack of the vanes helps to impart greater centrifugal force. The reduction in open area given the existence of the vanes tends to accelerate the gas speed. When scrubbing liquid is administered to this zone, impaction and shearing forces are applied as well. The spinning action of the gas tends to throw the liquid–gas mixture toward the vessel wall, where the distance between the droplets created and the gas is reduced. This action helps increase particulate capture.

A popular and clever vertical cage type vane scrubber is made by Entoleter, Inc. (Hamden, CT). Figure 18.2 shows this vane cage as viewed from above. The gas stream enters tangentially, and centrifugal force throws the larger

Clean Gas Outlet

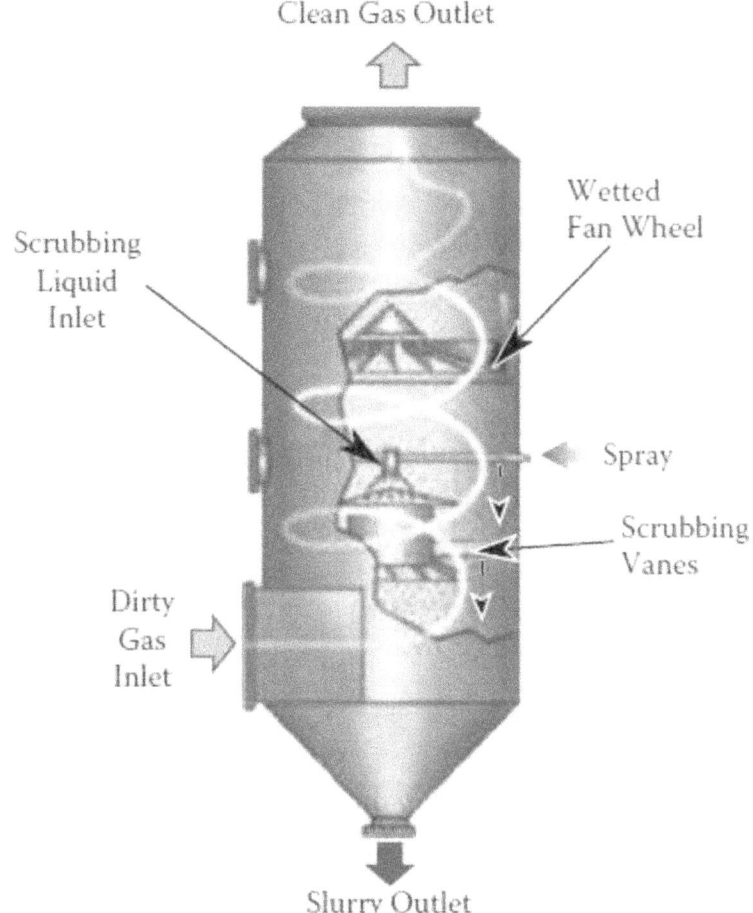

Wetted
Fan Wheel

Scrubbing
Liquid
Inlet

Spray

Scrubbing
Vanes

Dirty
Gas
Inlet

Slurry Outlet

FIGURE 18.1
Mikro-Vane scrubber. (MikroPul)

particulate to the vessel wall, where it impacts and is flushed down to the sump. The gases then follow a decreasing radius until they reach the vane cage. The slots in the vane cage function as a multiplicity of Venturi scrubber throats. When scrubbing liquid is injected into this swirling stream, the tangential motion tends to spin the liquid at an angle back outward into the path of the gas stream. This action increases the relative velocity between the gas and liquid and improves impaction and separation.

A spray cloud is formed in the vane cage zone, as diagrammed in Figure 18.3. You can see that the spinning action tries to throw the liquid outward, but the gas is being directed inward by the vanes. A droplet cloud is thus formed, and these droplets serve as targets for particulate capture much as in a Venturi scrubber.

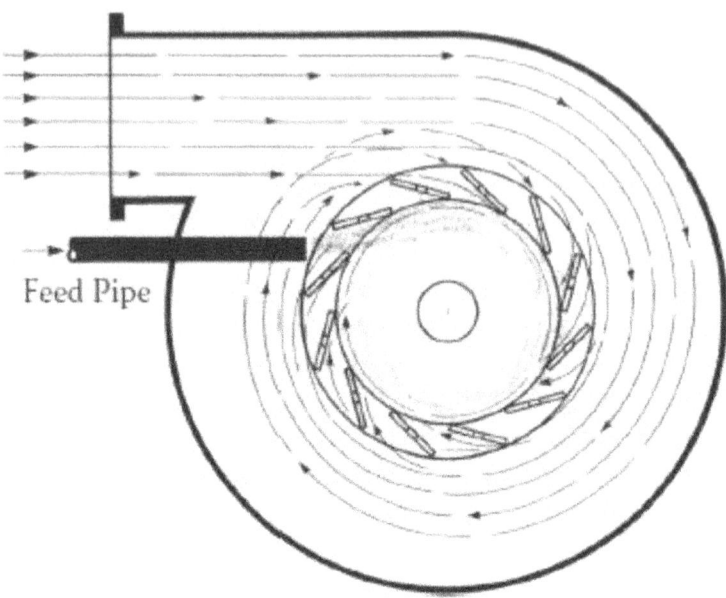

FIGURE 18.2
Centrifield scrubber. (Entoleter, Inc.)

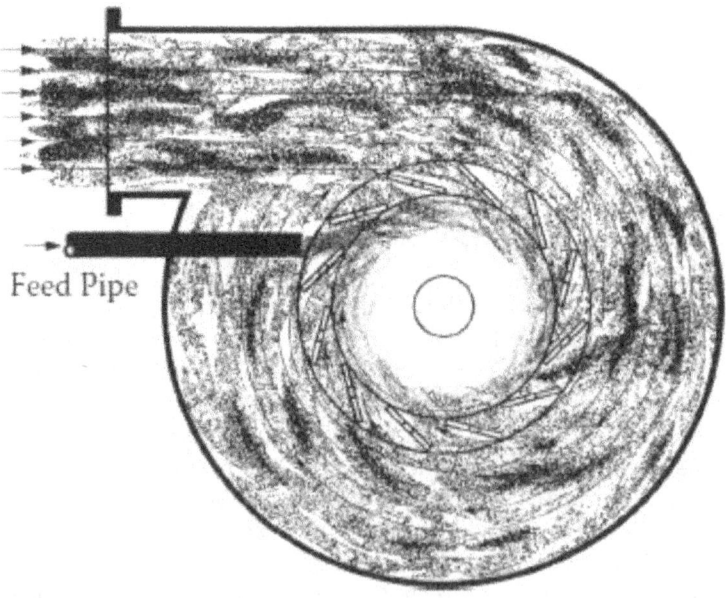

FIGURE 18.3
Centrifield cloud zone. (Entoleter, Inc.)

FIGURE 18.4
Vane type scrubber installation. (Entoleter, Inc.)

Vane type scrubbers are often found on rotary and tray type dryers such as found in the grain drying industry. Their compact size makes them attractive for roof mounting if space is a problem. As seen in Figure 18.4, a pair of vane type scrubbers are seen roof mounted with the exhaust stacks mounted directly on top of the scrubber's gas outlet.

When these types of scrubbers have a relatively short gas–liquid contact time, their gas absorption can suffer. To improve gas absorption, a vane type scrubber can often be combined with a gas absorber such as a packed tower as seen in Figure 18.5. A ring and cone trap is used to separate the two stages. The cone can be seen just below the packing section. This type of design can also be used to separate 10 μm particulate in the lower stage, followed by gas absorption in the upper stage, where both contaminants are present at the same time.

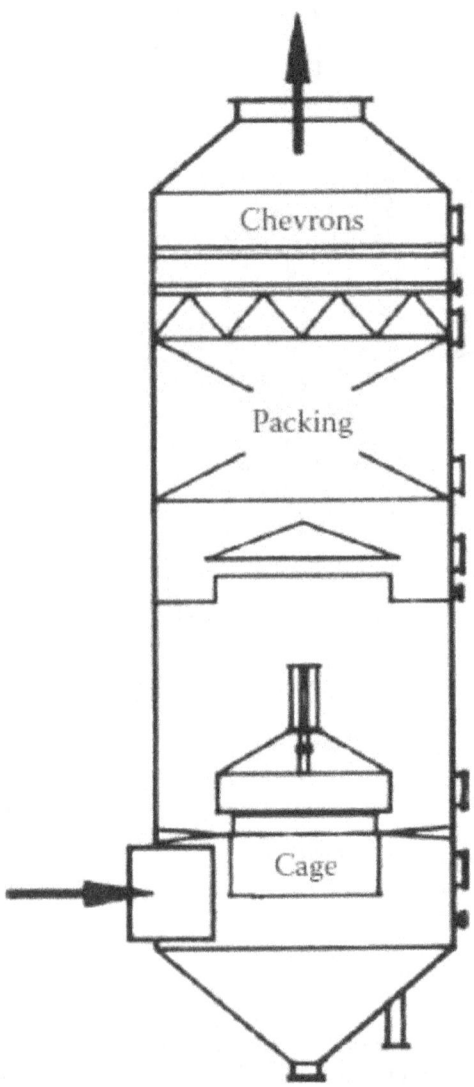

FIGURE 18.5
Vane scrubber plus packed bed. (Entoleter, Inc.)

Figure 18.6 shows a vane or centrifugal type scrubber in fabrication. The turbine-like vanes can clearly be seen. Note also that the central sleeve from which the vanes radiate. Many vane type scrubbers use a center disc or sleeve to act as a vortex finder that stabilizes the rotating gas pattern. When the vane is used as a droplet eliminator on larger units, this center sleeve often does double duty as an access manway to permit passage above or below the vane deck.

FIGURE 18.6
Vane type scrubber in fabrication. (TREMA Verfahrenstechnik GmbH)

18.5 Design Basics

Vane type scrubbers can operate at pressure drops of less than 3–4 inches water column to over 35 inches water column for the vertical cage type designs. They generally make good use of the scrubbing liquid and operate at low liquid-to-gas (L/G) ratios of 2 to 10 gallons/1000 actual cubic feet per minute (acfm) treated. Higher L/G ratios are used where the dust loading exceeds approximately 3–5 grs/dscf.

The vertical vessel velocities are like other cyclonic type wet scrubbers, that is, about 8–12 ft/s vertical velocity. Gas inlet speeds range from 35 to 55 ft/sec, but these speeds are often increased once inside the vessel to provide first-stage cyclonic separation. Gas outlet speeds are 40–55 ft/s if no stack is used and approximately 30–40 ft/s if a stack is in place.

These scrubbers usually require a disengaging space above the vane area because the gas stream is spinning, and the droplets require some time to separate. This disengaging zone varies by manufacturer but is typically ½ to 1 vessel diameter above the vane area. More disengaging zone is needed where horizontally oriented vanes are used because the gas flow tends to take an upward angular spiral rather than a flatter spin as in the vane cage type.

18.6 Operating Suggestions

Vane type scrubbers encompass several proprietary designs. It is therefore best to consult with the vendor regarding each specific application.

If the gas stream contains over 20% submicron particulate, vane type scrubbers may not be able to meet current air emissions regulations. The scrubber vendor should be able to predict the scrubber performance from a particle size analysis. When in doubt, perform or acquire an aerodynamic diameter particle size analysis for your application and submit it to the scrubber vendor for review. Given sufficiently large particulate, a vane type scrubber can provide economic performance versus medium- to low-energy competition such as Venturi scrubbers. Savings can accrue from reduced scrubbing liquid requirements and a more compact installation. Given the extensive use of internals in these designs, the capital cost may be higher than competitive designs because more material and fabrication labor time may be needed.

Given the generally low L/G ratios at which these designs operate, liquid distribution is critical. If spray distributors are used, strainers are recommended to reduce nozzle plugging. Upon internal inspection, care should be taken to observe the spray impact patterns and adjust the nozzle spray patterns or angles to provide complete liquid coverage. Telltale patterns can usually be seen on the vanes.

Because the vanes are inside the vessel, they can be attacked by corrosion from both sides. When selecting materials of construction, one should consider that the vanes should have sufficient thickness for double-sided corrosion. Too often, the vanes are thin, and localized attack can shorten their effective life. If the application is corrosive, remember that any vanes inside can be attacked from both sides, therefore your corrosion allowance should be doubled for interior components.

If the scrubber uses a lower-stage primary cyclonic knock-out section with central drain fitting, make certain that the scrubber is equipped with vortex breakers to stop the liquid from spinning so that the liquid may drain smoothly.

19

Venturi Scrubbers

19.1 Device Type

Venturi scrubbers are wet scrubbers that use a change in gas velocity to shear liquid streams (usually water) into tiny target droplets into which particulate and soluble gases are transferred. They are considered as a workhorse of the available air pollution control technologies given their low capital cost, reliability, and effectiveness on a variety of applications. They tend to use more energy than alternative designs, particularly on applications treating over 50,000 acfm of gases. Venturi scrubbers are used where the collected product can be handled wet. They are often used on processes, such as calciners and dryers, wherein the blowdown from the scrubber can be returned to a wet portion of the process. They can also handle the heavy-dust loadings, which can occur from these sources. Venturi scrubbers can ingest dust loadings of over 30 grs/dscf (grains per dry standard cubic foot) if designed correctly. Figure 19.1 shows a rectangular throat Venturi scrubber, a workhorse of the wet scrubbing industry.

19.2 Typical Applications

Venturi scrubbers are best used to remove particulate 0.6 μm aerodynamic diameter and larger where the gas flow is from 1 to 500,000 acfm if the particles are 10 μm and larger, and from 1 to 50,000 acfm if the particles are 0.6 μm and larger. They have been successfully used, however, to remove submicron particulate at pressure drops of up to about 60 inches water column.

If the gas stream has primarily submicron particulate (say, from a hazardous waste incinerator), a condensing wet scrubbing system, or a wet electrostatic precipitator, or similar lower-energy input system might be used instead.

There are literally hundreds of applications, however, in which the particulate is 1–20 μm diameter where the Venturi scrubber provides excellent results. The result is that thousands of Venturi scrubbers are in daily use throughout the world.

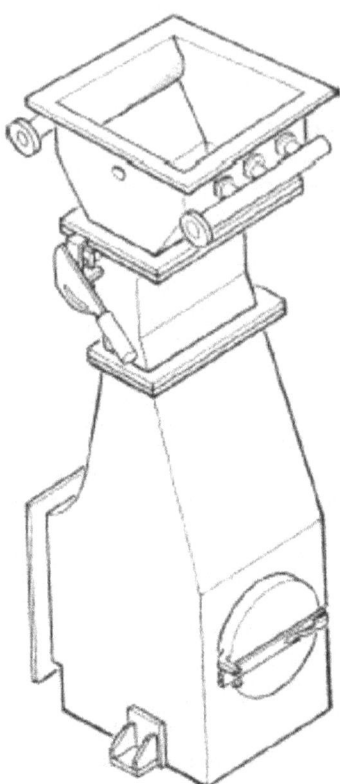

FIGURE 19.1
Venturi scrubber. (Bionomic Industries, Inc.)

Rectangular throat Venturis are commonly used on product dryers and calciners where there is a wet stage. Mineral lime kilns and lime sludge kilns (such as in the recausticizing section of a kraft pulp mill) often use Venturi scrubbers. Agricultural product rotary dryers are often equipped with primary product-collecting cyclones, which are followed by Venturi scrubbers. Grinding, milling (wet), mulling, and other operations that generate dust often use Venturi scrubbers for particulate control. Venturis on mineral lime kilns usually operate a 10–16-inch water column pressure drop, and units on lime sludge kilns are designed to run at 22–26 inches water column and sometimes higher if the lime mud being burned is high in sodium.

Boilers such as those firing bagasse or bark are often equipped with Venturi scrubbers. The boiler usually incorporates a primary knockout zone and cyclone collector followed by a medium-energy Venturi (approximately 10–15 inches water column).

Some metallurgic furnaces are equipped with higher energy Venturi scrubbers because the particles generated are smaller.

Annular Venturi scrubbers are used when the gas volume exceeds about 25,000 acfm. The reason for this is that designers like to maintain a throat width of 4–6 inches maximum. Sometimes a rectangular throat of this size would be too long to suit the gas inlet. The throat is therefore wrapped around to form the annular type. These designs are often seen on waste-burning boilers, larger kilns, and calciners, and large capacity dryers. Figure 19.2 shows an annular Venturi scrubber designed and built by TREMA in Europe. Note the ring-shaped liquid header at the top and the throat positioner at the bottom.

Eductor Venturi scrubbers are used where the designer wants to eliminate the use of a fan and is willing to use more liquid at higher pressure instead. These conditions might prevail where space is limited, the source may be explosive (a fan wheel spark could cause a problem), or the application requires simplicity. Eductors are used on tank vent systems, on tools used in the manufacture of semiconductor products, on odor control systems where

FIGURE 19.2
Annular Venturi scrubber. (TREMA Verfahrenstechnik GmbH)

fan noise may be an issue, and on emergency gas control systems (such as for chlorine control).

Reverse jet scrubbers are used on these same applications with primary focus on applications where the total energy input is an issue. They use a lower static pressure fan but a higher pressure pump, but they have a lower total energy input in many cases.

19.3 Operating Principles

All Venturi scrubbers operate by creating a dispersion of tightly packed target droplets into which the contaminant particulate is impacted. The droplet dispersion may be created by a high differential velocity between the scrubbing liquid and the gas resulting in a droplet-forming shearing effect. Other designs use pump hydraulic pressure and spray nozzles to generate the droplets. The overall intent is to impact the smaller particle into the larger droplet, which is more easily separated from the carrying gas stream inertially.

Once the particulate is impacted into the droplet, the droplet is separated from the gas stream using centrifugal force or interception on a waveform (chevron), baffle, or similar device.

19.4 Primary Mechanisms Used

Impaction is the primary collection mechanism in Venturi scrubbers (see Chapter 1). Interception and diffusion also come into play particularly at pressure drops above 10–15 inches water column where the smaller droplet size and droplet proximity enhance such capture mechanisms.

For gas absorption, diffusion is the primary method of capture. Venturi scrubbers can sometimes achieve 0.5–1.0 transfer units, although the residence time in the Venturi throat is truly short (typically milliseconds).

19.5 Design Basics

Typical Venturi scrubber types are as follows:

1. Rectangular throat designs, both fixed throat and adjustable.
2. Annular-type designs wherein the throat zone is an annular gap. This gap can be adjusted by moving the center body plumb-bob up and down to vary the open area and, therefore, the pressure drop.

3. Eductor Venturis wherein the momentum of pressurized liquid introduced into the device provides both mass transfer and motive force to the gas.

4. Reverse jet designs wherein the liquid is injected countercurrent to the gas flow. These designs force the particle into a nearly head-on collision with the liquid spray to enhance the application of the spray energy.

5. Collision-type designs split the gas streams and impact them nearly head-on to enhance momentum transfer from gas to particle.

6. Some Venturi scrubbers are made from parallel tubes or pipes as in the multi-Venturi (see Figure 19.9). These pipes may be oriented horizontally, vertically, or on an inclined angle. The scrubbing liquid is usually sprayed on the tubes or pipes. The slots formed between the pipes form the Venturi shape.

Gas inlet velocities for all these designs are generally the same as the ductwork conveying velocities, that is, 45–60 ft/s. The Venturi section outlet duct is usually sized for a similar velocity to reduce pressure losses through velocity changes.

The liquid rate for gas velocity atomized Venturis (using fans) is 5–30 gpm/1000 acfm treated, with 5–10 gallons/1000 acfm being common. The liquid-to-gas (L/G) ratio is increased as the inlet dust loading is increased. Liquid pressures are under 15 pounds per square inch (psig), with 5–10 psig being common. Hydraulically pressurized (spray nozzle type) Venturi scrubbers may use lower liquid rates, however it is the dust loading that truly dictates the liquid rate. The greater the particulate loading, the higher the liquid rate. Lime kilns, with inlet dust loadings of over 20 grs/dscf, may use 15–20 gallons/1000 acfm, whereas a dryer equipped with a product recovery cyclone may use only 4–8 gallons/1000 acfm. Figure 19.3 shows the way the L/G increases with increasing dust loading.

Various researchers have derived equations based on fluid mechanics to predict the pressure drop of a Venturi scrubber. Formulas by Howard Hesketh were presented in the book *Wet Scrubbers* (Technomic/CRC Publishers) and in other publications. Seymour Calvert, Shui-Chow Yung, and others produced useful equations that also predict the pressure drop. Venturi scrubber vendors use these predictions (often with some modification to suit their designs) to size the Venturi throat zone. It is therefore suggested that vendors be relied on to make Venturi throat parameter selections.

Suspended solids contents of 6%–8% and higher are not uncommon, although many units operate at 2%–4% suspended solids. This is significantly higher than many other wet scrubber designs (such as tray scrubbers). Designs using nozzles are typically limited to approximately 2%–4% suspended solids, otherwise nozzle plugging can occur.

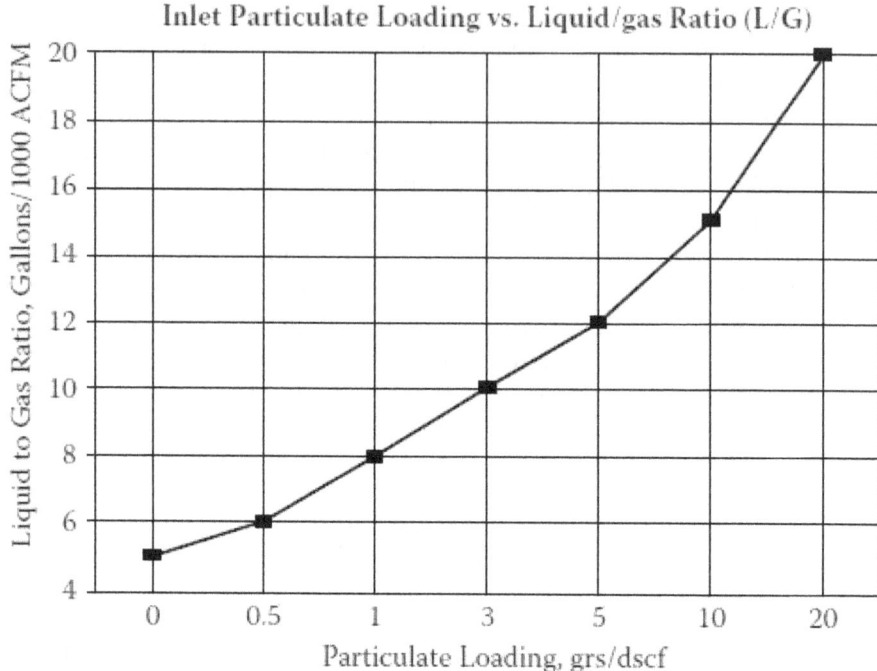

FIGURE 19.3
Liquid-to-gas (L/G) ratio versus loading.

Eductor type Venturi designs operate at much higher liquid rates and pressures because the liquid is also being used to create a draft. These units run at 20–50 gallons/1000 acfm, with header pressures of 30–60 psig being common.

Reverse jet designs have liquid rates in the range between the gas velocity atomized designs and the eductors. The liquid rate can be 50–100 gallons/1000 acfm or as low as 3–4 gallons/1000 acfm, depending on the dust loading and application.

Throat velocities vary from 70 to 90 ft/s to over 400 ft/s in high-energy designs.

Cyclonic separator vertical velocities range from 8 ft/s to 10 to 12 ft/s on larger systems (separators over 9–10 ft diameter).

The removal efficiency of a Venturi scrubber is a function of its pressure drop. Vendors have developed pressure drop versus efficiency curves as shown in Figure 19.4. Knowing the aerodynamic diameter of the particle (as determined by a cascade impactor), the designer can select the pressure drop at which the Venturi must operate. Often, removal guarantees can be provided based on only a known particle size distribution.

Let us look at various Venturi scrubber designs. A rendering in cutaway of an annular Venturi is shown in Figure 19.5. The gas inlet in this sketch is

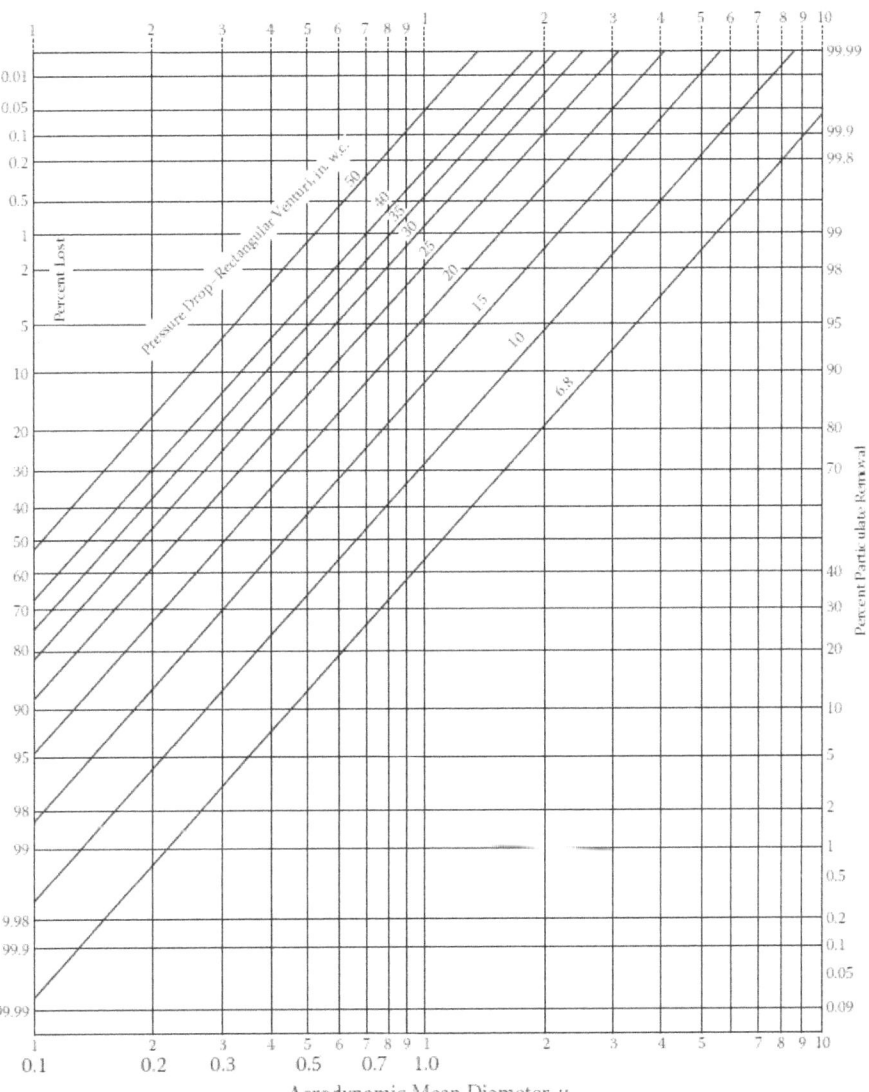

FIGURE 19.4
Composite fractional efficiency curve. (From Schifftner, K. and Hesketh, H., *Wet Scrubbers*, 2nd ed., Technomic Publishers, Lancaster, PA, 1996.)

at the top and the gas outlet is at the lower left. The conical device in the cut-away portion is the plumb bob. It defines the annular gap between itself and the tapered vessel wall. The slope or pitch angle of the plumb bob allows the throat area to be adjusted as the plumb bob moves up (to increase pressure drop) or down (to decrease pressure drop). The actuation is usually accomplished by mounting the plumb bob on a pipe, resulting in what looks like an

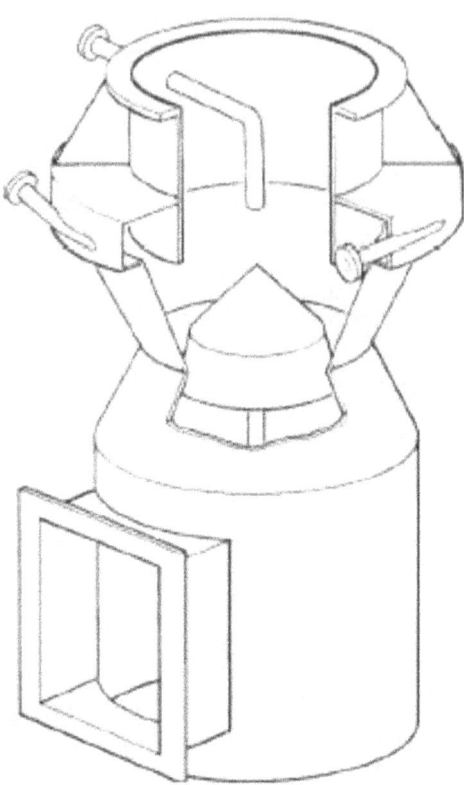

FIGURE 19.5
Annular Venturi. (Bionomic Industries, Inc.)

umbrella. The pipe extends down to the base of the Venturi and terminates outside the vessel. Moving this pipe or shaft up or down moves the plumb bob. A packed seal is incorporated surrounding the shaft to prevent leakage. These throats can be automated by using an electric or pneumatic jackscrew positioner to move the pipe based on pressure drop or draft signal.

Eductors, shown in Figure 19.6, operate by administering a jet of liquid (usually water) into the throat zone in the direction of gas travel. An energy exchange occurs between the liquid and gas. The high velocity and therefore kinetic energy of the liquid is exchanged with the surrounding gas, accelerating the gas. In part, the gas is also entrapped between droplet arrays and is pulled through the unit. The diverging section helps enhance the effect by allowing the droplets to slow down and achieve greater energy transfer.

Eductors can produce a draft at the eductor inlet without the use of an external gas-moving device (such as a fan). They are therefore often used where a rotating device such as a fan would not be compatible with the process or where space does not allow its installation. They are often used for small gas flows such as ventilating tanks or collecting dopant gases from

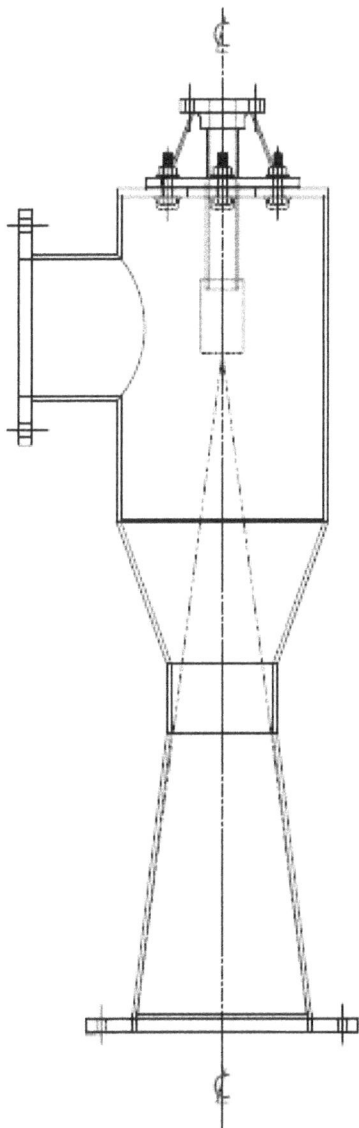

FIGURE 19.6
Eductor type Venturi. (Bionomic Industries, Inc.)

semiconductor manufacturing. The mechanical efficiency is quite low, however, so they are not commonly used on high volume applications (over 5000 acfm) without a supplemental fan.

The Dyna-Wave scrubber (Figure 19.7) improves impaction by spraying the scrubbing liquid countercurrent into the gas stream. The velocity of the liquid is directed into the gas stream, so the differential velocity is much higher

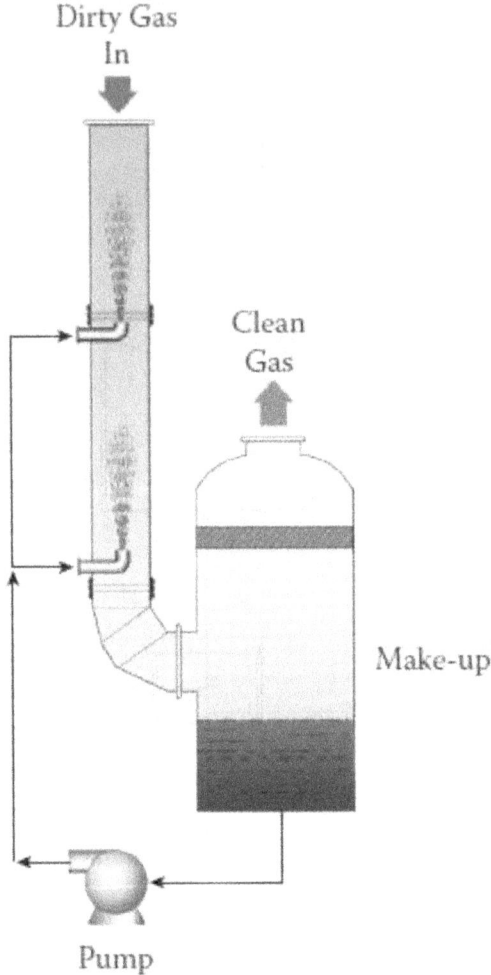

FIGURE 19.7
Reverse jet or Dyna-Wave Venturi. (Monsanto Enviro-Chem Systems, Inc.)

than in a conventional Venturi scrubber. This allows less gas side pressure drop to be used and can save horsepower by shifting the energy input duty from the low-efficiency fan to the higher efficiency pump.

A froth is created where the liquid reaches zero velocity and then turns 180° and moves concurrent with the gas. The particulate in the gas stream is impacted directly into this froth zone and is removed. Dyna-Wave scrubbers have been used on many particulate scrubbing applications. The resulting concurrent discharge of the liquid limits, to some extent, their gas absorption capability. In those cases, they are used in stages or are combined with absorbers such as packed towers.

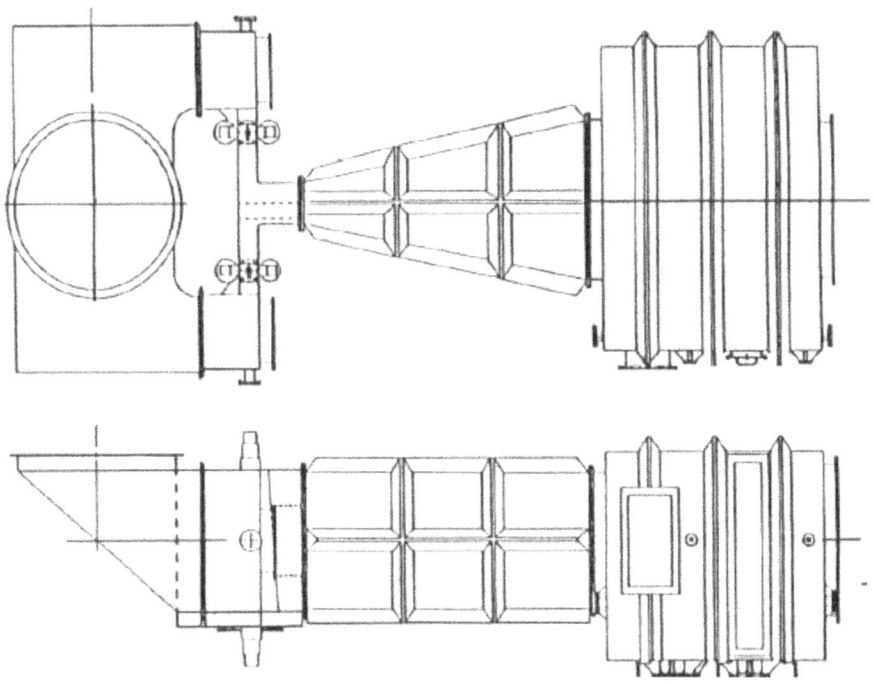

FIGURE 19.8
Collision scrubber. (Monsanto Enviro-Chem Systems, Inc.)

The collision scrubber shown in Figure 19.8 was developed by Seymour Calvert and has been used to collect submicron fumes from hazardous waste incinerators and other difficult applications. In this case, the inlet gas stream is split into two equal streams, turned 90°, and impacted head on. As in the Dyna-Wave, the goal is to maximize the differential in speed between the particle carried by the gas and the liquid. These Venturis can also be made to be adjustable using a movable T section mounted where the two throats converge.

The multi-Venturi shown in Figure 19.9 uses closely spaced rods or pipes that create long Venturi slots. It is known that an excessive throat width in a Venturi scrubber can result in a loss in efficiency. For that reason, and others, multiple Venturis are used. The throat width is reduced to a group of narrow slots. Although the total open throat area is nearly the same as in a conventional Venturi, the throat width is but a fraction of its conventional cousin. The wetted surface of the multi-Venturi is also greater. Some say that the increased wetted surface improves particulate removal. It does increase the cost, however, particularly if exotic alloys are used in its construction.

For all the designs, a separating device is used after the Venturi to remove the droplets that are now carrying the collected particles and absorbed gases.

FIGURE 19.9
Multi-Venturi (BACT Process Systems, Inc.)

Multi Venturi (BACT Engineering).

A cyclonic separator as shown in Figure 19.10 is a quite common application. Centrifugal force is used to spin the liquid droplets from the gas stream. Sometimes a packed tower or mesh pad type separator follows the Venturi (this is common for eductors, which may precede or be followed by packed towers for enhanced gas absorption).

Crossflow type droplet eliminators as shown in Figure 19.11 are also used. These use waveform type droplet eliminators (chevrons) that provide a surface upon which the droplets impact, accumulate, and drain. If the Venturi operates at over 35 inches water column, many vendors like to use crossflow droplet eliminators rather than cyclonic designs because the former offers greater small droplet removal than the latter. Without proper droplet control, the liquid could be entrained to the stack testing equipment and the particulate those droplets contain be counted as emissions. Droplet separation is critical.

In this configuration, the multi-venturi is located to the right and the crossflow separator is located to the left.

The crossflow separator typically uses modules configured in parallel such as described in Figure 19.11. The droplets impact on the chevron vanes and then drain downward out of the gas stream. Cleaning sprays are often used to wash the vane surface.

When a multiple-venturi design is in crossflow orientation, the droplet eliminator is integrated with the venturi stage. This results in a compact arrangement that may be useful in applications where headroom is a factor of the installation.

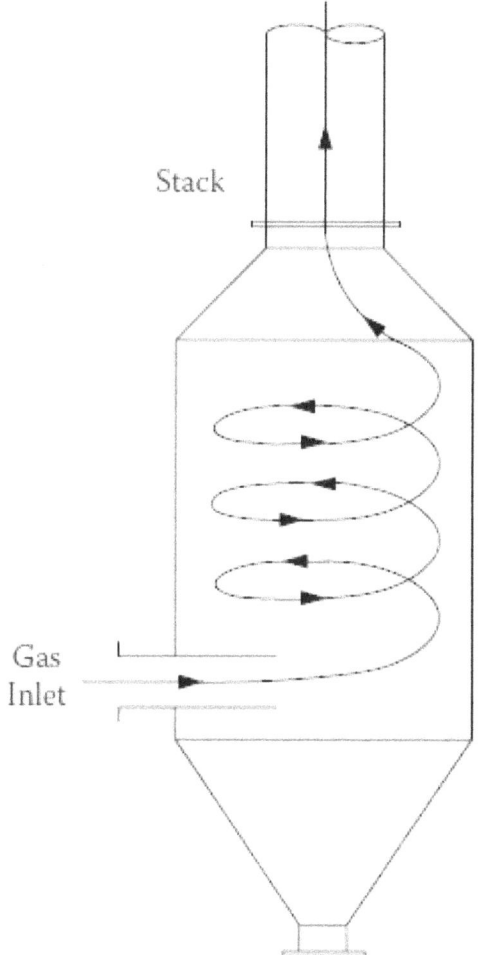

Stack

Gas
Inlet

FIGURE 19.10
Cyclonic separator. (Bionomic Industries, Inc.)

19.6 Operating/Application Suggestions

There are literally thousands of Venturi scrubbers in operation worldwide. General tricks of the trade include sending the scrubbing liquid to an elevation above the Venturi and letting the liquid drain from the bottom of the header into the Venturi if high solids loadings must be handled. Adjustable throats are of great benefit in setting the scrubber pressure drop and tuning the scrubber to the source. These adjustable throats are sometimes automated with a feedback loop to a differential pressure or draft controller that

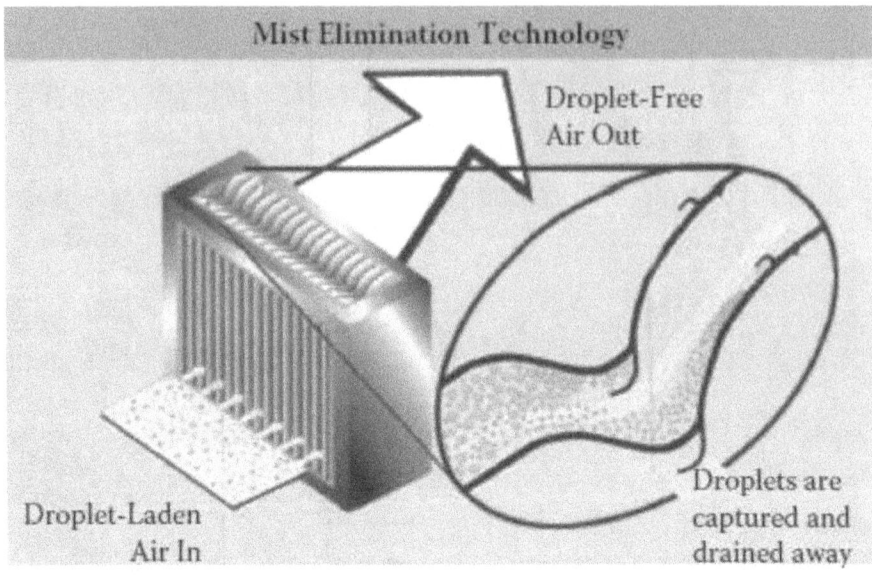

FIGURE 19.11
Crossflow droplet eliminator. (Munters Corp.)

allows the pressure drop to follow a process setpoint or an emissions permit parameter.

If the gas stream contains abrasive particles, wear plates are often used in the upper section (approach section), in the throat, and in the elbow area where the gases turn 90° to enter the separator. These elbows may also be

FIGURE 19.11
Multi-Venturi with Crossflow Separator (BACT Process Systems, Inc.

designed to be flooded with water, that is, the flooded elbow uses the water surface as an abrasion-resistant barrier. The conventional elbow is called a sweep elbow because it sweeps the gases toward the separator.

So-called horizontal Venturi scrubbers are usually inclined on an angle to allow liquid drainage. The gas and liquid tend to take a downward arc trajectory that limits performance, so horizontal Venturis are rare.

Separators are sometimes mounted on top of open surface decant tanks on applications where the collected product may float (such as bark char, bagasse fines, carbon black, etc.). Other units are operated "water-once-through" to flush high dust loadings to a remotely mounted clarifier.

Other systems use a product recovery liquid cyclone in the scrubber recirculation loop. These are sometimes used on foundry cupola or precious metals recovery applications. The underflow from the cyclone is sent to product recovery and the clears go to the Venturi headers.

If the recycle liquid contains solids (such as limestone), the liquid distributor to the Venturi is often mounted above the injection point so that the solids are flushed out of the bottom of the header, thereby reducing buildup and plugging. For clean liquids, the headers often discharge from the top so that the header is always full, and the liquid is evenly distributed.

The simple configuration and reliability of the Venturi scrubber makes it a true air pollution control workhorse.

20

Wet Electrostatic Precipitators*

Wayne T. Hartshorn

Hart Environmental, Inc., Lehighton, Pennsylvania

20.1 Device Type

The wet electrostatic precipitator (WESP) is a mechanical device that uses primarily electrostatic forces to separate particulate from gas streams. The collecting surfaces are periodically cleaned using water or other suitable conductive flushing liquid, thus the name wet electrostatic precipitator.

The basic components of a WESP are shown in Figure 20.1. They consist of either a low-level (shown) or high-level gas inlet, collecting tubes, mast-type electrodes mounted on a grid or frame, a high-voltage insulator section, an air-purged insulator compartment to prevent particulate from coating the high-voltage insulator section, a high-voltage power supply (transformer/rectifier set), and a gas outlet.

The designs also include various types of cleaning or irrigation systems that are used to purge the tubes of captured particulate. These purge systems may include fog nozzles, spray nozzles, or weir-type irrigation systems.

20.2 Typical Applications and Uses

WESPs are frequently used to collect submicron particulate that arises from combustion, drying operations, process chemical production, and similar sources. They are also used as polishing devices to reduce particulate loadings to extremely low levels. They are generally used where the inlet loading of particulate is under 0.5 grs/dscf (grains per dry standard cubic foot) and where corrosive gases may be present. They also excel where the particulate is sticky but can be water flushed. They often replace fiberbed filters or similar coalescing devices where solid particulate is present that could plug the fiberbed design.

---

* Additional information provided by Jerry Childress of McGill AirClean and David Meier of Bionomic Industries.

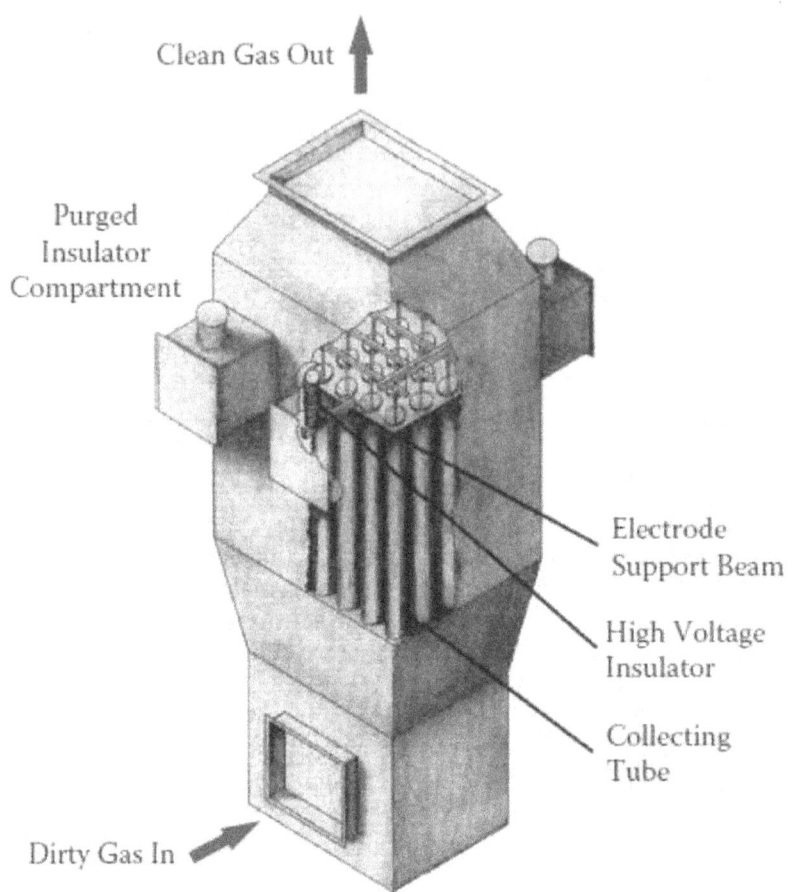

Clean Gas Out

Purged
Insulator
Compartment

Electrode
Support Beam

High Voltage
Insulator

Collecting
Tube

Dirty Gas In

FIGURE 20.1
WESP components. (Entoleter, Inc.)

Wet precipitators are increasingly being used as final cleanup devices behind and in combination with other air pollution control devices. Applications include chemical and hazardous waste incinerators; hog fuel boilers; acid mists; steel mill applications; vapor-condensed organics; nonferrous metal oxide fumes from calciners, roasters, and reverb furnaces; phosphate rock; veneer dryers; sludge incinerators; and blue haze and fume control. Figure 20.2 shows a WESP on a popular application, a veneer dryer.

The WESP can provide, in addition to fine or submicron particulate control, a final cleanup of mist elimination.

Another common application is on particle board dryers. These emissions can contain a combination of large particulate fines plus condensable aerosols. These products tend to be sticky, so the WESP, properly designed, is a good candidate for its control. On this unit, the WESP is in the center of the

FIGURE 20.2
WESP on veneer dryer. (Geoenergy International Corp.)

picture, and a droplet eliminator and fan is to the left of center. The gas flow is downward, thereby flushing solids toward the sump, assisted by gravity. The bypass stack for the dryer can be seen in the background.

Some of these applications require versatility. Figure 20.3 shows a WESP as applied to a combination fuel-fired boiler burning wood waste as the

FIGURE 20.3
WESP on combination fuel-fired boiler. (photo courtesy of McGill AirClean)

primary fuel. The other fuels may at any time be fuel oil, coal, natural gas, or noncondensable gases. The emissions arrive at about 450°F; therefore, a quench-type gas inlet is used. In this specific project, a gas flow of about 337,000 acfm was split between the two WESPs. The system controls particulate emissions to meet a strict opacity limitation. When gas volumes are high, the low-pressure drop of a WESP can be attractive in that the low-flow resistance saves energy at the prime mover (in this case, fans).

20.3 Primary Mechanisms Used

Electrostatic forces as well as diffusional forces are used to accomplish the separation. On some designs wherein the collecting tubes or surfaces are air or liquid cooled, thermophoretic forces are also used. In general, a series of zones are created wherein electrostatic forces sweep the particulate from the gas stream toward the contact (collecting) surface, which is periodically flushed with water to prevent the buildup of a resistive layer. One such application is shown in Figure 20.4. The WESP is in use on a particle board dryer.

To a minor extent, the WESP is also a gas absorber. The flushing system can also provide some mass transfer of contaminant gases into the liquid.

FIGURE 20.4
Particle board dryer WESP. (Geoenergy International Corp.)

20.4 Design Basics

WESPs consist of emitting electrodes mounted inside collecting tubes. A high voltage is introduced to the emitting electrode and a corona (charged field) is produced between the emitting electrode and the collecting electrode. Pollutant particles (sometimes solids, sometimes aerosols, often a mixture of both) pass through this corona and are moved toward the collecting electrode where they momentarily attach. Periodically, a flush of liquid (usually water) flushes the particulate away.

Many manufacturers have extended and extrapolated methods of sizing electrostatic precipitators. However, there has not been significant change in the state-of-the-art of electrostatic precipitation. Concentration has been centered on hardware improvements for reliability (Figure 20.5); voltage, and spark controls to maintain maximum stable electrical fields (Figure 20.6); increasing sizes to secure compliance with new and more stringent regulations; and attention to new and improved materials of construction for longer life and more resistance to corrosive gases (Figure 20.7). Further development work has resulted in more effective arrangements and configurations of collection and charging zones in the devices (Figure 20.8). Some of this work has provided for higher particle charging or more intense ionization

FIGURE 20.5
Electrode support of WESP. (Hart Environmental, Inc.)

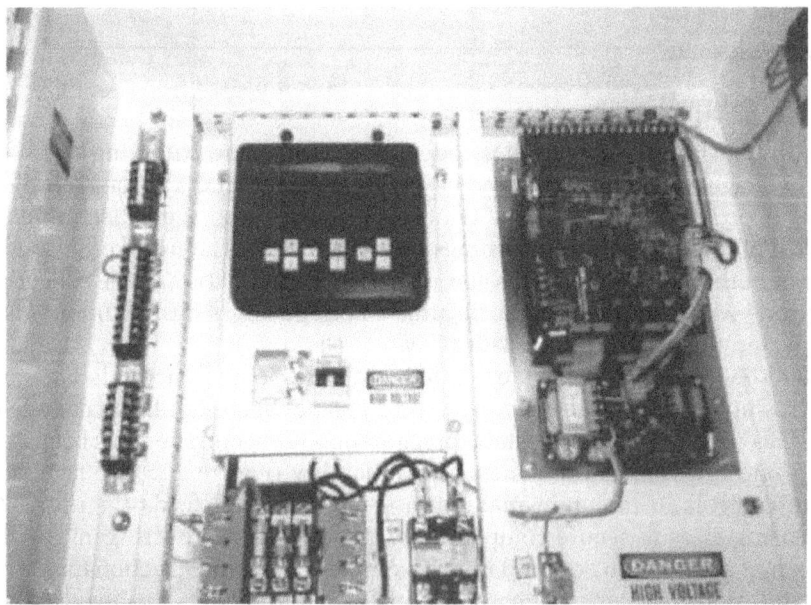

FIGURE 20.6
Modern WESP high-voltage controls. (Hart Environmental, Inc. Installation/NWL Control Corp.)

FIGURE 20.7
Picture of sonic development WESP designed and serviced by Wayne T. Hartshorn.

FIGURE 20.8
All-alloy WESP electrode bank. (Hart Environmental, Inc.)

FIGURE 20.9
Multiple discs on electrode. (Hart Environmental, Inc.)

(Figure 20.9). This has definitely added improvements to the state-of-the-art of fine particle collection.

Notice the insulators on either side of the discharge electrode mast (center), which passes through to the electrode frame located below.

To control the WESP and reduce sparking, modern solid-state controls are used that incorporate feedback-type logic. They bring the voltage up to the sparking potential then back off slightly, automatically, although the conditions in the WESP may vary.

The vertical tubular arrangement of the collecting tubes is shown in Figure 20.7. These tubes may be round or multisided, depending on the vendor.

To keep the discharge electrode masts centered, some firms use frames top and bottom. Modern designs use specially designed swivels that allow alignment of the electrodes and then lock them in place. These swivels are shown in Figure 20.8 just below the cross members. Because a WESP often handles corrosive gases, the vessel can be made from corrosion-resistant alloys or even nonmetallic fiberglass (if the surface is suitably prepared with a conducting surface).

To produce high efficiency, some vendors use multiple emitting discs on the discharge electrodes. These discs are shown in Figure 20.9 as they extend down into the collecting tube.

Discs are used instead of wire so that a series of intense corona fields can be produced. This can best be seen diagrammatically in Figure 20.10. The use

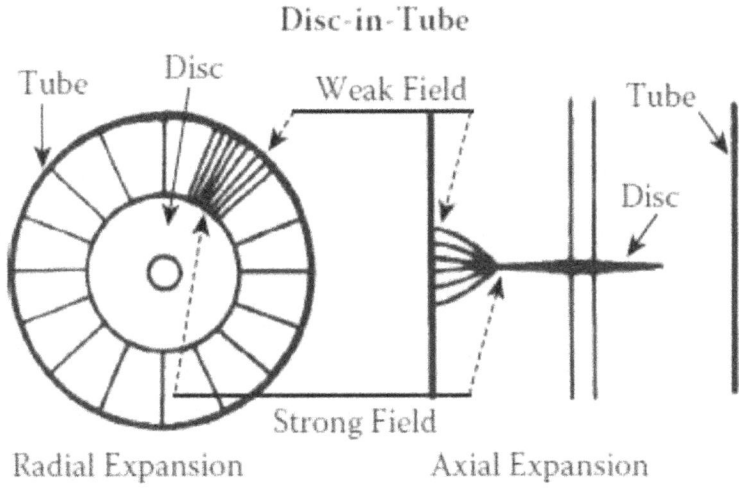

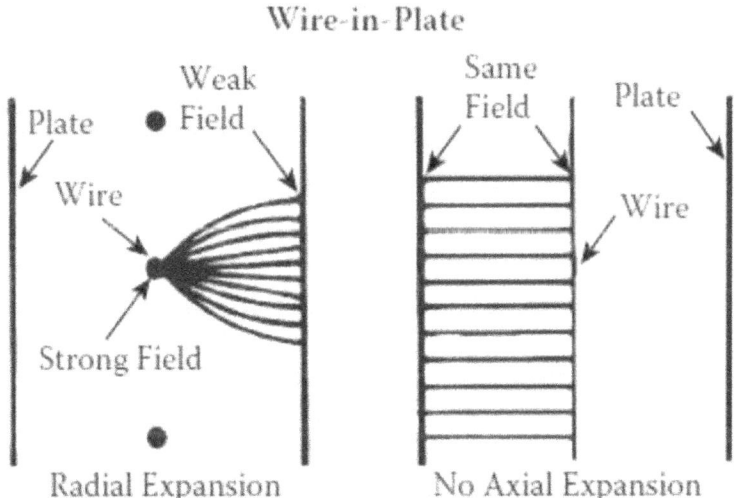

FIGURE 20.10
Disc versus wire corona formation comparison. (TurboSonic Technologies, Inc.)

of modern sparking controls has allowed the use of multiple discs and therefore multiple corona zones to be produced. A strong corona field can be produced between the edge of the disc and the collecting tube, much like the electrode to ground on an automotive spark plug. The controls of the WESP, however, allow a corona to be formed before the spark jumps the gap. This combination produces the greatest particulate control efficiency.

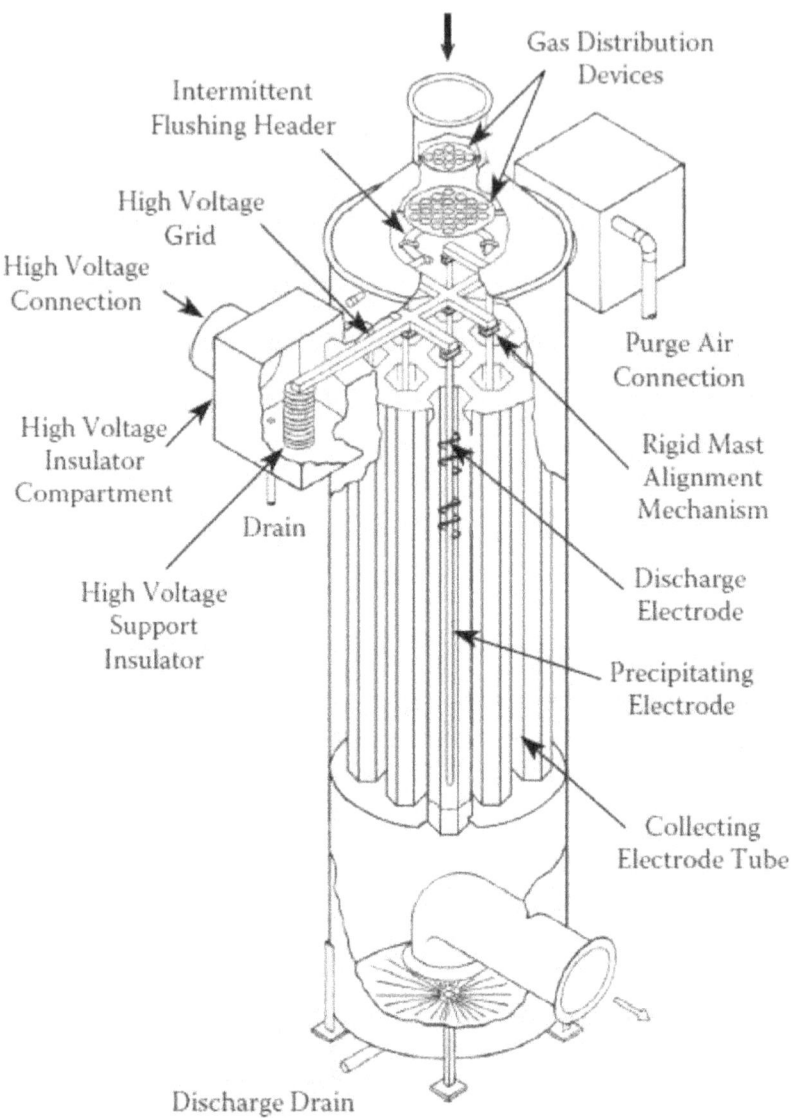

FIGURE 20.11
Basic components of a WESP. (TurboSonic Technologies, Inc.)

There are two types of electrostatic precipitator technologies. There is the dry electrostatic precipitator, which is cleaned of collected material by means of rapping and/or vibrating mechanisms. The wet precipitator is cleaned of collected material by means of irrigated collecting surfaces (Figure 20.11).

Until recently, the wet precipitators comprised a small share of the market for electrostatic precipitators. Originally, the leading application for wet precipitators was the collection of sulfuric acid. A typical unit was self-irrigating, tube type, and lead-lined fabrication. Reinforced thermosetting plastic has gained increased acceptance as well.

20.5 Types of Wet Precipitators

The design of WESPs can be characterized by configuration, arrangement, irrigating method, and materials of construction. Alloy construction of the entire WESP or just the emitting electrodes and tube bundle is common.

20.5.1 Configuration

There are two basic precipitator configurations: plate and tube. The plate type consists of parallel plates with discharge elements assembled between each plate. The tube type consists of an array of tubes, round or multisided, with a discharge electrode located in the center of each.

20.5.2 Arrangement

Gas flow can be arranged in parallel or series and horizontally or vertically. This feature also distinguishes a wet from a dry precipitator—because particles are removed from the latter through rapping, it is always arranged horizontally.

20.5.3 Irrigation Method

This has a greater impact on the operation of a wet precipitator than any other factor. There are many irrigation methods.

In self-irrigation, the most common method, captured liquid droplets wet the collecting surface. This method works only when the particles are mostly liquid. In a specialized variation, condensation from the gas stream wets the collecting surfaces. A cold fluid, usually air, is circulated on the outside of the collecting tube to promote condensation. As with mist collectors, irrigation by condensation works best with a gas stream high in moisture content and low in particle concentration. For this reason and others, the WESP is often used as a very high-efficiency mist eliminator after other gas cleaning devices such as fluidized beds and Venturi scrubbers. As shown in Figure 20.12, it can also be used after gas absorber/coolers such as packed towers wherein gases are cooled and then subcooled to condense water vapor onto water droplets (flux force condensation).

In spray irrigation, spray nozzles continuously irrigate the collecting surfaces. The spray droplets and the particles form the irrigating film. In intermittently flushed irrigation, the precipitator operates cyclically. During collection, it operates as a dry precipitator without rapping. It is periodically

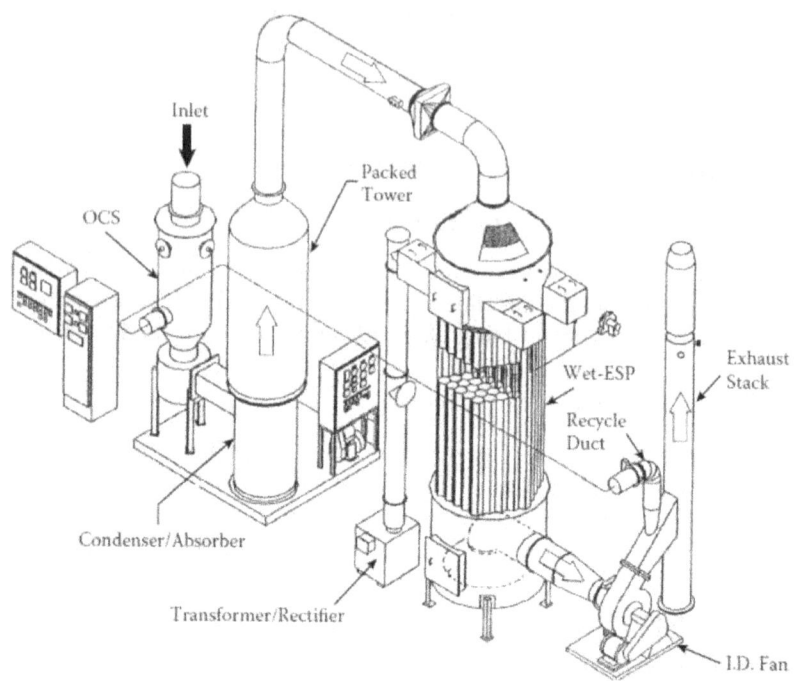

FIGURE 20.12
Flux force condensation type system with WESP. (TurboSonic Technologies, Inc.)

flushed by overhead spray nozzles. This method works well only if the particles are easily removed.

In film irrigation, a continuous liquid film flushes the collecting surface. Because the film also acts as the collecting surface, the plate or tube does nothing more than support the film. Therefore, the electrical conductivity of the irrigating fluid becomes an important factor. Nonconductive irrigants will not work. Also important are the physical properties of the film and the liquid-distribution network. The film must be smooth and well distributed to avoid high-voltage arcing, which can damage the unit and result in poor performance. Additionally, the distribution piping, plenums, and weirs must be designed to avoid dead zones that promote settling or plugging.

Electrostatic precipitation is made possible by the corona discharge. Through an effect known as the avalanche process, the corona discharge provides a simple and stable means of generating the ions to electrically charge and collect suspended particles or mists. In the avalanche process, gases in the vicinity of a negatively charged surface break down to form a plasma, or glow, region when the imposed voltage reaches a critical level (Figure 20.13). Free electrons in this region are then repulsed toward the positive, or grounded, surface and finally collide with gas molecules to form negative ions.

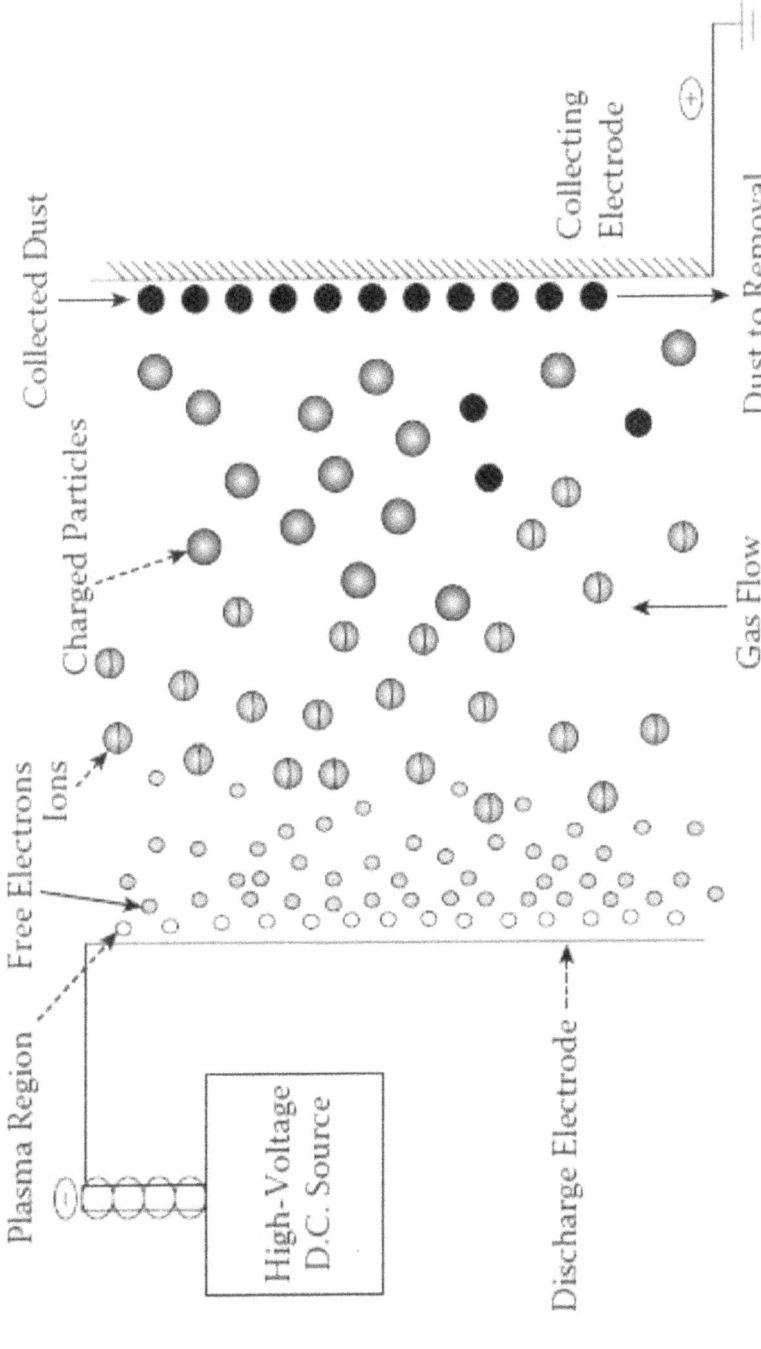

FIGURE 20.13
Electrostatic basics. (Wayne T. Hartshorn)

These ions, being of lower mobility, form a space-charge cloud of the same polarity as the emitting surface. By restricting further emission of high-speed electrons, the space charge tends to stabilize the corona. With a corona established, dust particles or mists in the area become charged by the ions present and are driven to the positive electrode by the electric field. Of course, for the foregoing to be successful, the proper electrode geometry, gas composition, and voltage must be present.

Particle charging is only the first step in the precipitation process. Once charged, the particles must be collected. As explained, this happens as a matter of course because the same forces that cause a particle to acquire a charge also drive the like-polarity particle to the grounded surface.

The next step is particle removal. In a wet precipitator, the material is rinsed from the collecting surface with an irrigating liquid.

20.6 Selecting a Wet Electrostatic Precipitator

The Deutsch equation describes precipitator efficiency under conditions of turbulent flow:

$$E = 1 - \exp(-AW/Q)$$

where:

E = collection efficiency, 1 − (outlet particle concentration/inlet particle concentration)

A = area of the collecting surface

W = velocity of particle migration to the collecting surface

Q = upward gas flow rate (gas velocity × cross-sectional area of the passage)

The derivation of the equation depends on simplifying assumptions, the most important being that all particles are the same size, the gas velocity profile is uniform, a captured particle stays captured, the electric field is uniform, and no zones are untreated.

To account for the numerous variables, a modified Deutsch equation is used, in which the term W (particle migration velocity) is replaced by another known as *effective migration velocity* (EMV). Empirically determined, EMV is a characterizing parameter that accounts for all the nonidealities mentioned, as well as for the true particle-migration velocity. Values for EMV used in the modified form are considerably lower than true particle velocities calculated or measured in the laboratory.

Most WESPs do not suffer from the nonidealities encountered by the dry-type devices. Also, because the wet-type precipitator is frequently configured for vertical gas flow, sneakby is avoided. Therefore, EMV values for wet precipitators are usually higher than those for dry precipitators. This means that, for a specific application, a wet device can be smaller than an equivalent dry device. This is additionally true because a wet precipitator operates on a cooled, lower volume gas stream.

Because the collecting surfaces in a wet precipitator are cleaned by a liquid, the wet precipitator can be used for virtually any particle emission.

Generally, the physical and chemical properties of the particles are not an important factor in the design of wet precipitators, as well as factors that are normally of concern in the design of dry precipitators, such as electrical resistivity, surface adhesion, and flammability. A possible exception is the dielectric constant of the particles. It has a weak effect on the maximum charge that can be achieved, according to the theoretical relationship for predicting particle saturation charge.

$$N = \{1 + 2[(k-1)/(k+2)]\}(E_o \Sigma a^2 / e)$$

where:
- N = saturation charge
- K = dielectric constant
- E_o = charging field
- A = particle diameter
- E = electron charge

The effect of dielectric constant on performance is not normally considered in the design of precipitators because the dielectric constant of most particles is high and has little effect on the charge. However, the constant may be important in oil mist collection by a wet precipitator. Some oils tend to have very low constants, which can markedly lower collection efficiencies.

Nevertheless, there are many applications for which a wet precipitator should be carefully considered and even some for which wet precipitation should be the only technology of choice (Figure 20.14). Some such conditions occur when the gas stream has already been treated in a wet scrubber, the temperature of the gas stream is low and its moisture content is high, gas and particles must be simultaneously removed, the loading of submicron particles is high and removal must be very efficient, liquid particles are to be collected, and the dust to be collected is best handled in liquid.

Unlike other gas cleaning methods, the applicability of wet precipitators strongly depends on the particular design. In some cases, certain wet precipitator designs may not be suitable for certain applications. For instance, a precipitator for gas streams containing adherent particles must be continuously, not intermittently, irrigated.

The second most important factor in design after the type and configuration have been decided is materials of construction. Wet precipitators operate at, or below, the adiabatic saturation temperature of the irrigating fluid (usually water), and corrosion is a constant concern.

	Scrubber	Fabric Filter	Dry ESP	Wet Precipitator
Fine Particles		X	X	X
Liquid Particles	X			X
Low Gas Temp./ High Dew Point	X			X
Sticky Particulate	X			X
High Efficiency		X	X	X
Gas Absorb. Req'd	X			X
High Resistivity Particles	X	X		X

FIGURE 20.14
Application comparison chart. (Wayne T. Hartshorn)

Wet precipitators are rarely made of carbon steel, at least the surfaces that are in contact with the gases to be treated. Carbon steel construction may be feasible only when the gas stream is high in pH and low in oxygen. Ordinarily, wet precipitators are constructed of one or more corrosion-resistant materials. These materials can include simple stainless steels, exotic high-nickel alloys, reinforced thermo-setting materials, and thermoplastics.

From a materials standpoint, the casing, or housing, is the least critical element. The outside of the shell housing not in contact with the gases need not even be corrosion resistant, only capable of withstanding ambient conditions. The collecting surfaces should afford the maximum resistance to chemical attack. Also, fabrication points subject to corrosion should be minimized because failures in the collecting surfaces can disturb the electric field and cause arcing and lowering performance. Because the discharge electrodes are usually not irrigated, there is a concentrating effect on their surfaces that does not occur on wetted areas. For example, if the gas stream contains 200–500 ppm SO_2, 10–20 ppm HCl, and 0–5 ppm HF, the pH on the moist surface of the discharge electrodes will be about 1.0, even if the irrigant is kept at a pH of 3.0 or higher. The galvanic effect of operation in the range of 40,000-V direct current compounds the corrosion potential of the concentrating effect. For these reasons, the discharge electrodes should always be fabricated of a material of significantly greater corrosion resistance than that of any other part of the wet precipitator.

Wet precipitators capture fine or submicron particles without high-energy consumption (Figure 20.15). Their capture efficiency of submicron particles is greater than that of the highest energy wet scrubber. The size of the wet precipitator strongly affects its performance in collecting fine particles.

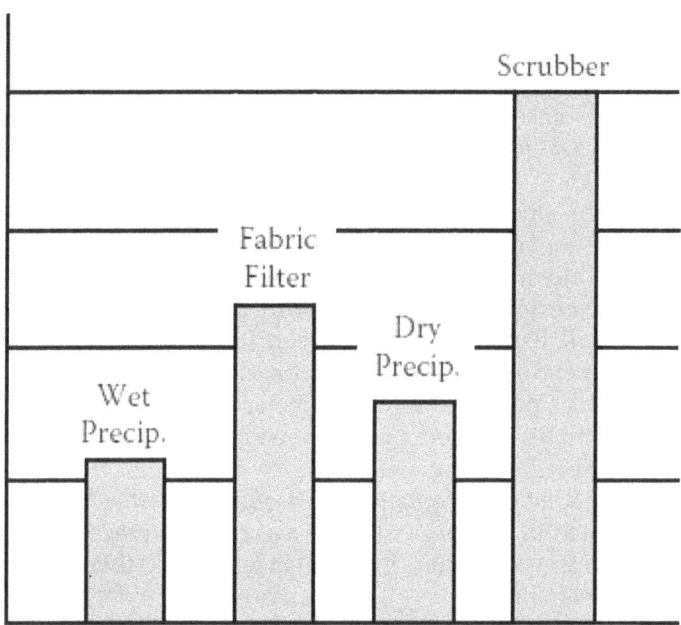

FIGURE 20.15
Relative energy consumption. (Hart Environmental, Inc.)

Wet precipitators are particularly effective in capturing large particles. However, most gas cleaners do a good job of this; 30%–40% of the emissions from a dry precipitator consist of large particles, mainly because of emissions due to rapping and reentrainment. Similarly, a considerable portion of the emissions from a wet scrubber is caused by mist carryover (another form of large particles).

20.7 Operating Suggestions

Wet precipitators are relatively insensitive to the chemical and physical characteristics of the gas stream or the particles. Gas streams at almost any temperature or of any composition can ultimately be treated with the proper design. With added quenching and conditioning, wet precipitators can handle flue gases at over 2000°F because the adiabatic saturation temperature will always be less than approximately 180°F. Because wet precipitators can be constructed from a wide variety of materials, they can treat the most aggressive gas streams.

The factors that most influence the cost of wet precipitators are collection efficiency requirements, materials of construction due to the corrosive nature of the gas stream, and physical size due to the gas volume to be treated.

The actual cost of a wet precipitator in most cases will be site specific. A cost and systems analysis should be performed to determine the configuration, materials of construction, and size. Typically, a wet precipitator system to treat corrosive gases can run from $100 to $300 per square foot of collecting surface area; for noncorrosive applications, the price may be in the $50–$90 range.

Wet precipitator operating costs are among the lowest for gas cleaning equipment. They operate at lower pressure drops than scrubbers or fabric filters and generally have less collecting area and require less high-voltage power than dry precipitators. For estimating purposes, high-voltage power consumption will usually range between 0.1 and 0.5 W/actual ft^3/min gas volume, depending on collection efficiency requirements. Auxiliary equipment, such as purge air blowers, heaters, and pumps, are highly site specific, so estimates of their power consumption should be done on a case-by-case basis.

Regarding installation orientation, it is suggested that the high-voltage supply be mounted in the serviceable area as close as practical to the WESP. This keeps the high-voltage runs minimal in length and therefore less expensive to install and maintain.

The WESP is a very effective device for use in the collection of submicron particulate and mists where those contaminants can be water flushed from the collecting surfaces.

21

Special Applications: Venturi "Scrubbers" as Evaporators

21.1 Device Type

Venturi type scrubbers have been used for decades for the direct contact evaporation of water from process liquids such as pulp mill black liquor. These devices were typically used in stages wherein the black liquor was gradually increased in concentration to avoid foaming issues and to provide adequate control over the solids. The Venturi design had advantages since the gas path is essentially open and thus reduces the chance of plugging.

The design has also been applied to waste incinerators wherein a zero or near-zero liquid discharge is required. Venturi scrubbers are used since they can recirculate at a higher solids content than most spray or tray type devices.

Of course, any wet scrubber that uses the direct contact of hot, dry gas with water will experience evaporation. As mentioned in Chapter 1, "Air Pollution Control 101," the proper application of a wet scrubber to a hot, dry gas source will result in that gas stream becoming saturated with water vapor. The source of that vapor is the evaporation of the water from the liquid being recycled. Fluidized bed type contactors can also be used as evaporators; however, the peak solids content that can be reached is typically less than that of a Venturi scrubber.

21.2 Typical Applications

Direct contact concentrators have been used to increase the solids content of landfill leachate, to produce water from natural gas recovery operations (such as in "fracking"), and to a lesser extent for process liquid stream concentration. The device is used effectively wherein direct contact of the liquid with hot gases is acceptable or where surface type evaporators plug with solids on the heat exchange surfaces.

In recent years at the time of this writing, interest has increased in evaporating excess water from liquid streams, particularly from those liquid streams relating to gas recovery and/or leachate treatment. Evaporating excess water reduces the volume and cost of trucking or pumping the liquid streams away for treatment off-site. If the treatment is on-site, removing much of the excess water can reduce the size and cost of the on-site treatment equipment. Sometimes, the evaporated water can be recovered through condensation if the process to which the system is attached needs make-up water.

21.3 Operating Principles

The direct contact concentrator works by mixing a hot gas source with a dispersion of droplets. The dispersion is typically produced using a Venturi scrubber that can operate at a high solids content rather than a spray nozzle type system. This technology was pioneered in the 1960s and 1970s in the direct contact evaporation of black liquor in the kraft paper industry. The key feature is evaporation using large droplets rather than small droplets. Why? Small droplets evaporate more quickly; however, the solids they contain can be spray dried into small particles that could result in a particulate air emission. Using large droplets is less efficient but reduces, if not eliminates, the fine particulate formation. The droplets are separated using chevron or cyclonic action as described in other chapters.

21.4 Primary Mechanism Used

The primary mechanism is evaporation. The evaporation is promoted and enhanced by increasing the surface area of the liquid and carrying away the evaporated water using a moving hot gas stream. When a liquid stream is mixed with a hot gas stream that is not saturated with water vapor, the "liquid" in the droplet heats up and changes phase to water vapor. In doing so, the solids in the liquid begin to concentrate. When the vapor surrounding the liquid becomes saturated with water vapor, the evaporation ceases. By constantly renewing the mix of unsaturated gases with the liquid, the evaporation can be caused to continue. At some point, however, the liquid solids content becomes so high that pumping and other practical conditions (such as erosion) limit the solids content to which the mixture can reach. The primary mechanism, however, is adiabatic saturation in the quencher and Venturi throat contact zones.

21.5 Design Basics

The direct contact concentrator is designed to evaporate water on a continuous basis from process wastewater streams. The typical system is designed to minimize the net energy input to accomplish the concentration and, if water recovery is needed, to reclaim to the greatest extent possible the evaporated water. In some instances, the recovered water can reduce the need for new infeed water in the process.

A photograph of a compact, transportable direct contact Venturi evaporator is shown in Figure 21.1. The major components consist of a hot gas inlet equipped with a quench section followed by a Venturi throat (often adjustable to vary the contact droplet size) and then a high-efficiency droplet separator. The hot gases are often pulled through the system using an induced draft fan.

We will look at a system that includes both evaporation and water recovery because it encompasses both stages. For evaporation only, the water vapor will exit directly to the atmosphere. The evaporation-only designs

FIGURE 21.1
Photograph of concentrator. (Bionomic Industries, Inc.)

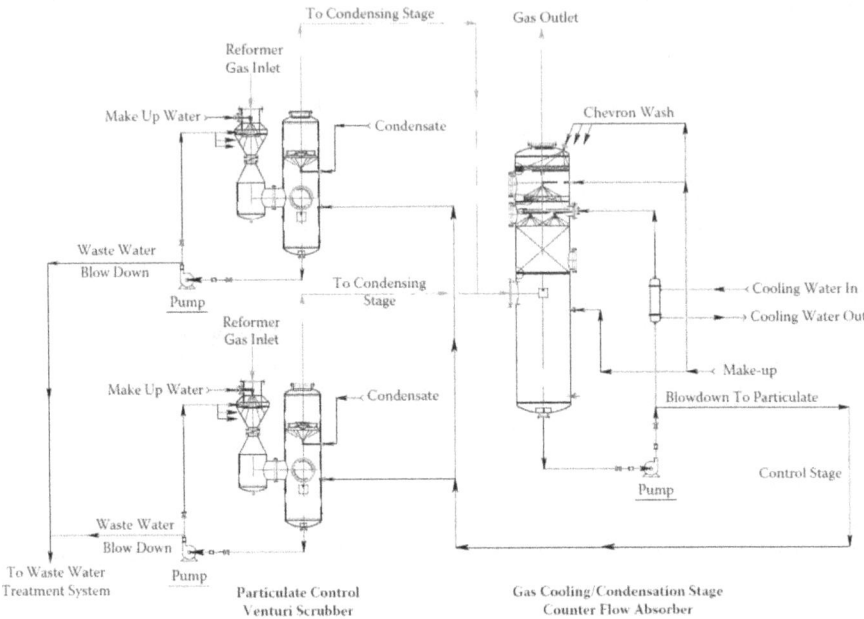

FIGURE 21.2
Component diagram. (Bionomic Industries, Inc.)

are characterized by their visible water vapor plumes. With water recovery, because the plume is essentially condensed, the plume is essentially eliminated. The water recovery stage is discussed in greater detail in the next chapter since, in condensing the water, heat can also be recovered. Figure 21.2 shows the primary component parts of such a system.

To make the system transportable, these evaporators are often skid mounted. All the components mentioned above are fitted onto a support skid. Figure 21.3 shows such a skid being moved in the fabrication shop just prior to shipment. The view angle is from the pump end, and the support skid can be clearly seen. If the components are too large to facilitate truck shipment, multiple units can be set side by side or subassemblies can be produced that house the larger components. The droplet separators, for example, may be too large for easy transport and thus could be shipped separately. Both horizontal separators and vertical cyclonic separators can be used. These systems are basically classic quencher/Venturi scrubbers applied to the specific task of evaporation.

The components for the system with condensate recovery, forced draft, include the following:

1. A heat source. This may be a burner combusting natural gas, syngas, producer gas, oil, biofuel, or other heat-producing liquid fuel. It could also be a solid-fuel burner. The heat input is typically in the range of 300 to 1800°F.

FIGURE 21.3
Skid-mounted Venturi evaporator. (Bionomic Industries, Inc.)

2. A gas inlet with safety bypass to atmosphere. The bypass allows the heat to be rejected safely in case of the heat source system malfunction. It is also used during burner purge for gas-fired burner startup.

3. A blower capable of overcoming the system gas-flow resistance. The blower is equipped with an inlet filter and, if necessary, a sound reduction device.

4. A Venturi scrubber (co-current flow) or counterflow tray or fluidized two-fluid contact device. The device brings the hot, dry gas stream into intimate contact with the liquid stream that requires concentration, resulting in the evaporation of moisture from that liquid. Since solids in the liquid are concentrated during the evaporation, the gas–liquid contact device must have a superior ability to operate at a high suspended and dissolved solids content.

5. A pump to draw the concentrated liquid from the sump of the contact device and recirculate the liquid so that the solids concentration will increase to a desired level.

6. An optional heat exchanger to preheat the process liquid to enhance the evaporation. The hot concentrate extracted from the bottom of the contact device exchanges part of its heat to preheat the infeed liquid.

7. A droplet control stage to separate the droplets created in the device from the gas stream. The droplet control helps keep the concentrated liquid in the device for the maximum evaporation cycles. If condensate recovery is not required, the gases would proceed to an induced draft fan and typically a stack. If condensate recovery is required, the following additional equipment is used.

8. A direct-contact condenser tower (such as a packed tower) that places cooled water into direct contact with the water-saturated gas stream, thus condensing and recovering water.

9. A heat exchanger to extract the heat from the liquid recycled in the direct-contact condenser.

10. A pump to recycle the recycled water and condensate through the heat exchanger and back to the direct-contact condenser.

11. A pump to recycle the cooling loop water to an air-cooled heat exchanger or heat sink (lake, pond, cold plant process stream exchanger, etc.).

12. A bleed line to bleed away the condensate.

13. An exhaust stack to vent the residual gases.

There are some disadvantages to the technique. Venturi type concentrators send the hot gases co-current with the liquid stream. This means the heated liquid passes in the same direction as the hot air and in doing so minimizes the difference in temperature between the liquid and the gas, which is known to reduce the evaporation rate. The infeed liquid is typically fed at the hottest point in the device (the inlet), wherein it dilutes the liquid that is being recycled and proceeds co-current with the gas stream. Adding the dilute liquid to the hottest zone can cause spray drying of solids from the liquid and a visible emission of particulate. Diluting with the infeed and allowing the infeed to move co-current with the concentrate limits the maximum solids concentration that the liquid can reach.

If the process liquid contains odorous compounds, those compounds could possibly strip from the liquid stream, thus requiring further treatment. The direct-contact technology works best in concentrating liquid streams that contain inert or dissolved solids but are devoid of low-solubility liquid or gases that can be stripped from the liquid.

21.6 Operating Suggestions

To achieve the best results, the direct-contact concentrator should be applied only to sources wherein concern for stripping of odors or volatile organic compounds is at a minimum; otherwise, these compounds could strip from the system. If such a liquid stream must be treated, the gases from the concentrator could be subsequently scrubbed (for example, using a neutralizing or oxidizing chemistry) with the blowdown from that stage being returned for concentration in the Venturi. As an alternative, if the vent gases do contain stripped compounds, it may be possible to send those gases to a thermal oxidizer and then use the resulting hot gases in the concentrator.

A density control is suggested to limit the recycle solids before their concentration can cause mechanical problems. It is not uncommon to require decanters or centrifugal separation devices to further increase the solids content.

If the water vapor is to be recovered, careful control of the cooling water to the condensing stage is suggested. An efficient and more precise way of doing this is to use a temperature-regulating control valve in the cooling water bypass around the heat exchanger (rather than varying pump speed).

22

Energy Recovery

22.1 Device Type

Packed towers, tray towers, and even spray towers can be used for energy recovery through the direct contact of a heat-absorbing liquid (usually water) with a hot gas stream. These devices may also simultaneously remove soluble gaseous pollutants. In these systems, the *rate* of heat transfer is controlled by the difference in temperature between the hot gas and absorbing liquid. The *amount* of heat absorbed is controlled by the mass of liquid that passes through the device. Devices that are strictly counterflow, such as a packed or tray tower, exhibit higher heat transfer efficiency than devices such as a spray tower that may entrain liquid upward and therefore reduce the differential in temperature required for optimum heat transfer. With proper design, however, spray towers can be effectively used for heat recovery.

22.2 Typical Applications and Uses

Most wet scrubbers applied to hot sources inherently transfer heat. As mentioned in other chapters, the initial heat transfer may be used to adiabatically saturate the gas stream. Thereafter, more heat can be removed through the sensible cooling of the gas stream. As the gases mix with the liquid, both the latent heat (the heat removed as the water vapor in the gas stream changes phase to liquid water) and the sensible heat are recovered. This is quite different from what occurs with surface (nondirect contact) type heat exchangers such as economizers or tubular heat exchangers. Those devices recover only sensible heat.

Given the preceding, the best application sources for direct-contact heat recovery therefore are those devices that emit high humidity (lots of condensable water vapor from which the latent heat can be recovered) and hot gases. Such sources are boiler flue gas stacks wherein the hot gases can first be adiabatically saturated, sources that emit lots of water vapor or steam (such as food processing or paper making), thermal oxidizers (particularly ones that oxidize wet fuels), steam vents, and similar hot wet sources.

The recovered heat may be used, for example, to preheat boiler feed water. It may be used to provide supplemental heat for a building or loading dock. In other applications, it can be used to thaw frozen products. More exotic uses harness the heat to drive electric generator systems, which in effect are like driving an air-conditioner backward. Sometimes the recovered heat is used to preheat the wastewater outfall from a plant so that further biological destruction can more favorably occur. In all of these, the economics must be investigated. As fossil fuel costs rise, the economics favor heat recovery. Conversely, low fossil fuel costs can make the use of heat recovery only marginally beneficial. Each application must be evaluated based upon its own merit (or lack thereof).

Within these applications are two basic types. In one, the goal is to recover the heat simply by direct contact with the liquid in the heat transfer tower. In another, the liquid circuit must be cleaner; thus, the recovered heat is subsequently transferred to a "clean" liquid stream. In the latter, a heat exchanger (liquid–liquid) is used to transfer the heat in the "dirty" heat recovery tower water to clean process water. The process water may itself be a heat transfer liquid (such as ethylene glycol) that will subsequently be used to indirectly heat a device (such as a jacketed process vessel or reactor).

Figure 22.1 shows the direct-contact heat recovery process wherein the tower water itself absorbs the heat and no further transfer is used—thus the

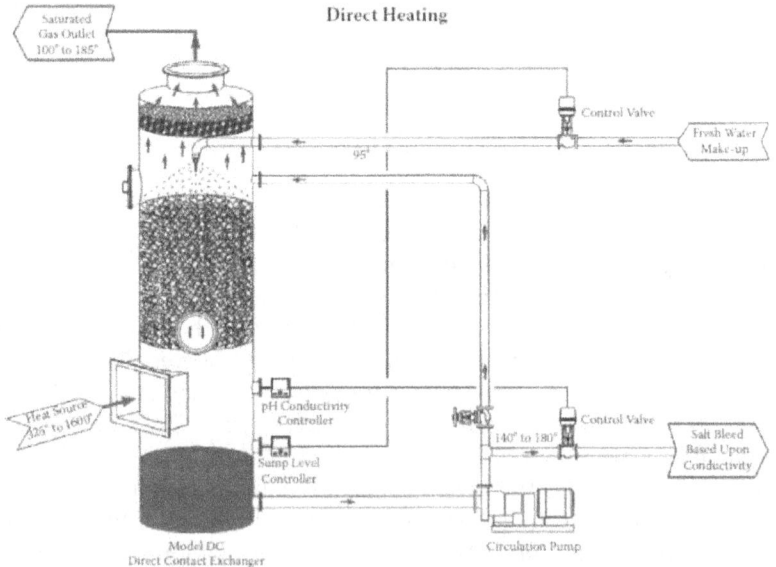

FIGURE 22.1
Heat recovery *direct* type diagram. (Bionomic Industries, Inc.)

term *direct*. This type of application may be one that heats water used for area heating or other applications wherein there are limited concerns that the tower water may mix with other liquid streams. The feed water simply enters the liquid distributor at the top of the tower, descends through a gas/liquid contact zone (in this case a packed tower), and then is drained away. The tower is usually sized "water once through," since recirculating the heated water to the top of the tower would reduce the thermal efficiency by decreasing the temperature differential between the water and the gas.

Notice the use of a direct contact counterflow packed tower as the heat transfer device. The liquid circuit can be recycled and bled (as shown) or once through as mentioned earlier. The tower design is like any packed tower. The packing media may be metallic, nonmetallic (plastics), or, quite often, ceramic packing, since these devices are connected to hot sources and ceramic packing resists overheat better than, particularly, plastic packing. (See Section 22.5, "Design Basics.")

Figure 22.2 shows a similar system, but the dirty water heat is transferred to a clean liquid using a heat exchanger. With these systems, it is still best to operate water once through; however, sometimes the water is recirculated and only the make-up water is added to the top of the tower. In the latter design, the amount of heat being recovered is more important than the temperature to which the liquid is heated. With a recirculation system, the overall

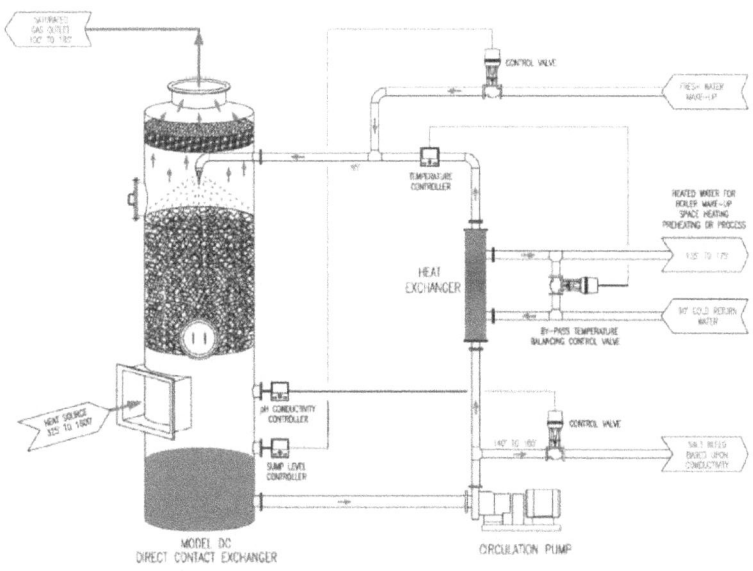

FIGURE 22.2
Heat recovery *indirect* type diagram. (Bionomic Industries, Inc.)

thermal efficiency is less, again given the reduced temperature differential in the tower, though in practice this lower efficiency is typically minor.

What if the hot gas source also contains acid gases and the like? Remember, the heat recovery tower is basically a packed, tray, or spray tower. These devices will also absorb soluble gases (as explained in other chapters). During the absorption, say, of acidic flue gas, heats of absorption are evolved. Add a reactive chemical such as caustic and heats of reaction are also evolved. These heats can also be recovered in the heat recovery tower.

In systems without heat recovery, the heats of absorption and reaction usually go overboard via heat carried away in the scrubber blowdown or in the water vapor leaving the stack or both. In systems with heat recovery, that heat is captured and perhaps can be used to offset the facilities use of expensive fossil fuels.

What about the water vapor plume? If the system did not use heat recovery, the scrubber may emit a dense water vapor plume. These plumes are quite visible, and the public often mistakes these plumes for pollution. With a heat recovery application, the plume is reduced if not eliminated. Often, only under rare atmospheric conditions (low temperature and high relative humidity) can a water vapor plume be seen.

What about water consumption? Some installations want to not only recover heat but also conserve water. The water that initially evaporates as the hot gases mix with the water in the heat recovery tower can often also be recovered. Indeed, the amount recovered typically must be balanced with the water pollution control system's capability at the specific site. In some applications wherein a high hydrocarbon-based fuel is oxidized, the combustion product water vapor is also recovered.

22.3 Operating Principles

Energy recovery devices are typically adaptations of packed, tray, or spray towers. The operating principles are described in Chapters 9, 10, 12, 14, and 17.

The difference with the heat recovery designs is that they are often configured to operate based upon the liquid rate rather than the gas rate. For example, a customer may want to heat a given amount of wastewater to a not-to-exceed temperature. The heat recovery tower would therefore be sized based primarily on that liquid rate and temperature. The hot source may offer more heat than is necessary, therefore, the tower may need only a slip stream from that source. The hot source heat production may be marginal, however, and the tower may be designed to recirculate, and bleed based upon the final (controlled) liquid temperature specified.

22.4 Primary Mechanism Used

For heat recovery, the mechanism is basic. Heat tends to move from hot sources to colder surfaces. In these devices, the hot "surface" is the incoming gas mixture, and the cold surface is a constantly replenished liquid surface (usually water). These towers are sized much like the soluble gas absorbers described in other chapters.

As mentioned earlier, the difference in temperature between the hot gas source and the liquid determines the rate of heat transfer. The greater the temperature difference (delta T), the greater the heat transfer rate. The amount of heat finally transferred, however, is controlled by the mass of liquid passing through the tower. In English units, the basic equation is as follows:

$$Q = m(cp)(\text{delta } T)$$

$$Q = \text{Total heat available}(BTU/hr)$$

$$m = \text{mass of liquid}(\text{usually water}), lbs/hr$$

$$(cp) = \text{liquid specific heat}(\text{if water, about } 1)$$

$$(\text{delta } t) = \text{liquid temperature, Out-In,} °F$$

In all these devices, the hot gases are brought into direct contact with the recirculated liquid in preferentially a counterflow mode.

22.5 Design Basics

Given that the source to which the device may be applied is hot, the towers often are designed for a thermal upset. If packed towers are used, the packing may be heat-resistant polypropylene or even ceramic saddles. A photo of the type of packing often used is shown in Figure 22.3. This general type of packing is often used as heat-transfer media in regenerative thermal oxidizers.

Given the extra weight per cubic foot of ceramic packing, the tower needs to be designed for this greater load, particularly at the packing support. Often, on larger diameter towers (more than 4–5 feet), an additional central beam or column is used to transfer the packing weight to the vessel wall or base.

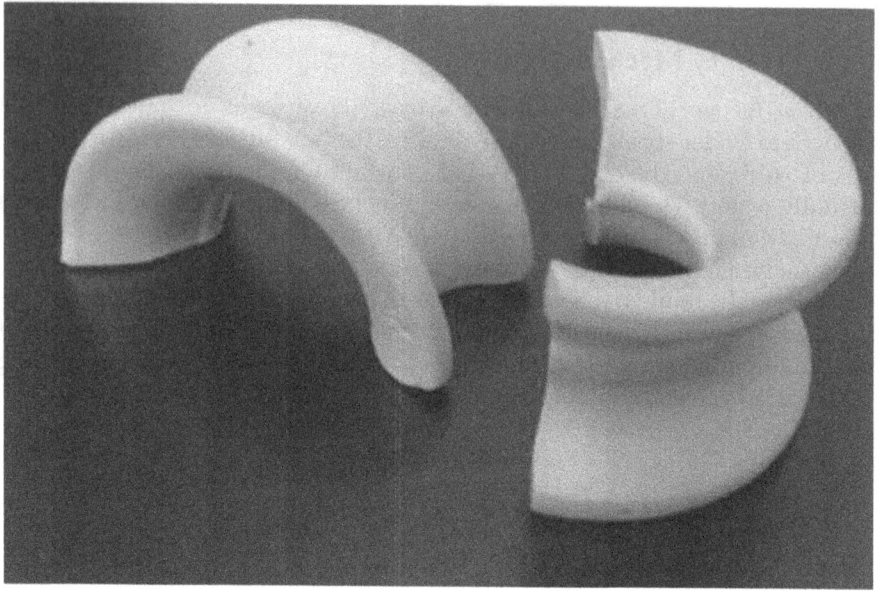

FIGURE 22.3
Ceramic/porcelain saddles. (Lantec Products, Inc.)

Ceramic packing also has wetting characteristics different from plastic or metal packing. The surface is lower per unit volume, and the liquid tends to be retained on the rougher surface of the packing rather than slide off. These parameters, however, have been well defined by the packing makers. The packing supplier is therefore consulted regarding the specific packing depth and water rate. Typical packing depths are 6–10 ft.

Vertical gas velocities for towers using ceramic packing range from about 4 to 6 ft/s. Vertical velocities for towers packed with heat-resistant plastic packing range from about 5 to 7 ft/s. The liquid (irrigation) rates per horizontal square foot are dictated by the type packing but range from about 8 g/ft² to about 12 g/ft². The irrigation rate ensures that the packing is fully wetted. Dry locations or areas that are not constantly wetted decrease the heat transfer rate. For liquid distribution, weir type or low-pressure (under 15 psig) spray type liquid distributors are commonly used.

Though the vessels are often fiberglass-reinforced plastic construction, sometimes stainless steel is used given the possibility for overheat. If stainless steel is used in a tray type design, stainless steel trays are also used.

For spray towers for heat recovery, multiple spray zones are used with a bias toward locating them higher in the tower so that the descending droplets can better mix with the hot gases. In addition, the droplet eliminator often receives any make-up (assumed cold) water so that the droplet eliminator acts as an additional mass-transfer surface.

22.6 Operating Suggestions

It is best to accumulate the maximum and operating conditions of the heat source both for its thermal (BTU) capacity and for the presence of any contaminant gases. Since these devices place the liquid in direct contact with the gas stream, both heat and contaminants can be passed to the liquid. If contaminant gases exist (for example, SO_2 from a fossil-fueled boiler source), the design can also include chemical neutralization of the absorbed contaminant. The system design would need to include chemical addition (i.e., pH control and conductivity control) equipment.

If the source does not include contaminants, basic temperature control can be used. A common method is to use a liquid outlet temperature sensor as a primary control point with sump level as a secondary control. Since liquid is typically evaporated from these systems and the make-up water (or liquid stream) is typically at a fixed rate, sometimes only a portion of the liquid flow is sent through the heat recovery unit. The flows can be recombined after the device, or that portion can be heated to a specific higher temperature.

Since the heat recovery demand may vary (say, seasonally), these systems often use a hot gas bypass (incorporating diverting dampers) so that the recovery unit is used only when needed.

If a fan is used to move gases through the unit, a bit more energy can sometimes be recovered if the fan is placed ahead of the heat recovery tower, thus recovering the heat of compression of the fan. In this case, the exhaust stack is often attached to the tower exhaust; thus, the tower becomes a stack base.

23

Multitechnique Equipment for Gasification (Syngas)

23.1 Device Type(s)

The term *gasification* has been applied to a variety of technologies that take a solid "fuel" (or furnish, such as wood waste, silage, waste grasses, pulp mill black liquor, etc.) and convert it to gases. Incinerators operating under starved-air or sub stoichiometric combustion systems are often called gasifiers because they indeed are—and the term is more politically correct than *incinerator*.

The fuel (typically a biomass furnish) is converted in the process to a gas with the minimal application of oxygen, and then the resulting gas is separately converted to a fuel gas or combusted, thus producing heat. These sources may be equipped with cyclone collectors and dry precipitators or even baghouses. They may also use wet Venturi scrubbers or wet electrostatic precipitators. The final gas stream may or may not be released to the atmosphere, depending upon the process.

In other gasifiers, the furnish is converted under pressure and temperature (often using steam) to produce a syngas that is further catalytically converted to a fuel gas, dimethyl ether, or other product. In effect, the syngas producer is like a large, continuous pressure cooker. The furnish is converted to a syngas, and the syngas is cleaned and catalytically converted to other products (typically using a Fisher-Tropsch process). In some of these systems, a small portion of the syngas produced must be sufficiently cleaned so that it can be used to produce the heat to drive the gasification process. Depending upon the type of combustor used, the syngas stream must be cleaned at exceedingly high removal efficiency. Since water vapor is often just "along for the ride" and impacts the system size and cost, it is often removed through direct-contact condensation techniques. The water vapor content may also affect the catalytic conversion operation; thus, the water vapor content is carefully controlled.

Modern "pressure cooker" reformer type syngas gas cleaning equipment is truly a hybrid design. These applications typically use dry cyclone

collectors for primary particulate separation followed by Venturi scrubbers for supplemental particulate separation and gas saturation, followed
by counterflow gas coolers to squeeze out water vapor. Some systems may
even include an additional wet scrubbing stage to remove sulfides (such as
hydrogen sulfide) that might foul the downstream catalyst. These pressure
cookers process gas cleaning stages are an inherent part of the process and
are usually not considered an emission control system (the treated gases do
not directly emit to the atmosphere); therefore, they come under the pur-
view of the gasification system designer rather that the regulatory body. The
specific design requirements are often set by the designer of the gasifier—
not by code.

Figure 23.1 is a picture of the gas cleaning portion of a gasification system
used to produce aviation-grade biofuel. Dry cyclones are shown toward
the top of the photo, and a Venturi scrubber and condenser are shown to
the left.

FIGURE 23.1
Gasification system, gas cleaning portion. (Bionomic Industries, Inc.)

23.2 Typical Applications and Uses

To reduce our demand for fossil fuel, systems are being designed to convert waste materials into energy-producing products. The waste material typically is biomass such as wood waste (cellulosic), silage, grasses, sludges, and higher sulfur coal that, if the coal were to be burned, would produce excess SO_2 emissions. The biomass fuel is often called the *furnish* for the process. The furnish is typically solid but can also be liquid (such as pulp mill black liquor) that contains lignin or other organics that can be converted to syngas.

Though an oversimplification, the gasifiers used can be divided into three basic groups: starved-air gasification, plasma destruction, and reformer gasification.

23.2.1 Starved-Air Gasification

The first is a gasifier that converts the furnish to syngas under starved-air (substoichiometric) conditions. Since minimal air is used, the resulting gas stream is typically high in carbon monoxide (CO) and low in CO_2. The gas mixture properties typically reflect a higher molecular weight and higher specific heat than standard air. The gasifier itself may be of fluidized bed design or be a specially modified boiler.

To control the emissions, dry cyclone collectors to control particulate followed by dry or wet electrostatic precipitators may be used. Dry cyclones followed by a wet Venturi scrubber could also be applied.

23.2.2 Plasma Destruction

Another method involves the application of an electric arc that produces a plasma that dissociates the furnish into component gases. Those gases are reformed into the syngas. These systems operate with little or no oxygen, and the gas stream is extremely high in hydrogen (H_2). The high H_2 content reduces the molecular weight and dramatically increases the specific heat of the resulting gas stream. These effects combine to elevate the saturation temperature of the gas mixture.

If the furnish to the plasma unit contains chlorinated components, the system can be designed to remove the hydrochloric acid (HCl) that is formed and to increase the acid concentration in the blowdown by using multiple stage packed or tray tower absorption. In that type of acid gas recovery system, a dilute stage water-only absorber is preceded by one or more higher concentration absorbers. Make-up water is introduced to the last stage, and the blowdown is bled forward to the next upstream stage until the blowdown reaches the first stage. The concentrated acid is then bled from the first stage. For HCl recovering, acid concentrations of up to about 18% wt. can be reached.

Figure 23.2 depicts the various components of a gas cleaning system as applied to a plasma type syngas gasifier. The technique is called *hybrid* because each stage uses a different technique to deal with the gas conditions at that stage. As the gas stream is cleaned, not only are pollutants removed but the gas is also cooled. Wet scrubbers have been successfully applied to plasma type systems. Dry cyclones may be used for primary particulate separation; however, the conveying velocities in these systems are so low that inert gases are usually removed through settling in the lower portion of the plasma unit.

23.2.3 Reformer Gasification

The third common method is the use of a reformer. The reformer uses steam to react with carbon to produce H_2 and CO when heat is applied.

$$H_2O + C + heat >>>>>> H_2 + CO$$

In addition, the water reacts with the CO to produce additional H_2.

$$CO + H_2O >>>> H_2 + CO_2$$

The heat comes from the separate combustion of a portion of the produced fuel gas. The gas cleaning system takes the raw reformer gas and cleans it of particulate that might adversely affect the combustion stage. The gas stream water vapor content is carefully controlled by adjusting the cooling temperature of the condensing stage of the gas cleaning system. In the reformer circuit, the CO_2 essentially goes along for the ride and is maintained at as low a concentration as practical. Since the CO_2 is formed via the water shift reaction with the water and not from combustion with air, the CO_2 content is low. In the combustion (heat) zone, the fuel is syngas that is high in H_2, so the stack CO_2 emissions are low.

In the reformer (or pressure cooker) systems, the operating pressure may be 10–15 psig up to 900 psig, depending upon the design. The furnish is mixed with pressurized steam and maintained under those conditions so that the H_2 will form. It is not unusual for the reformer to take days of such "cooking" before H_2 is produced. Thereafter, once stability is obtained, new furnish is added continually and inert material (ash, metallic oxides, etc.) is withdrawn. Since steam is used in the process, the gas stream characteristics at each stage must also be calculated since the extra water vapor in the gas density can vary greatly. The H_2 content is typically lower with a reformer type system than with a plasma type system, but the condensing demand is higher given the presence of the steam. The CO content as measured at the gas cleaning system inlet is often higher than that seen in other gasification systems. The heat exchange circuit becomes a heat recovery stage, which improves the overall thermal efficiency as well as acts to control the water vapor content for the shift of the CO to H_2 and carbon dioxide.

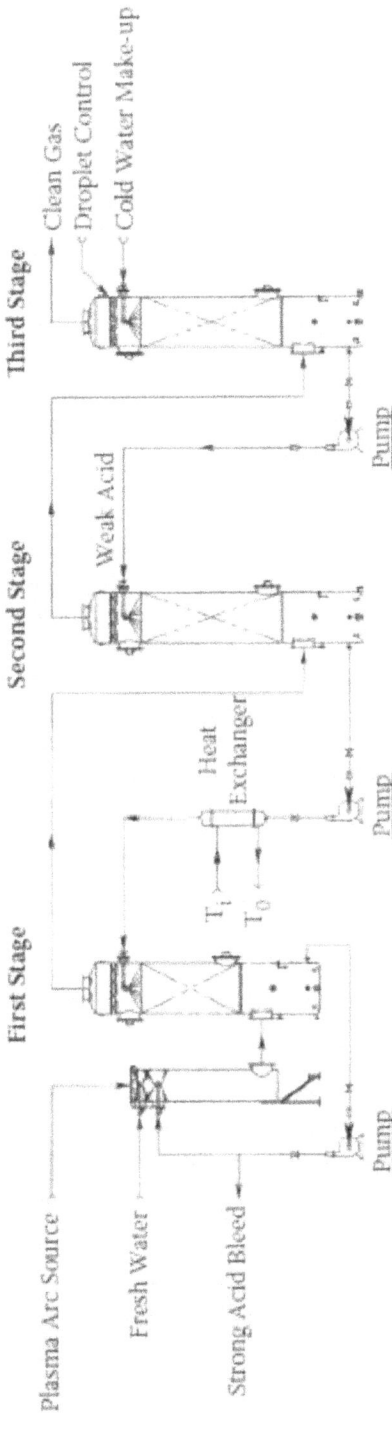

FIGURE 23.2
Plasma gasifier diagram. (Bionomic Industries, Inc.)

23.3 Operating Principles

As mentioned previously, the gasification process produces gases of high specific heat and low molecular weight. The contaminants can include finely divided particulate, tars, and other condensable and hydrocarbons such as phenols and aldehydes. No single gas cleaning technique can remove all those contaminants.

Currently, the typical system includes the use of dry cyclones for primary particulate removal while the gas stream is hottest, followed by Venturi scrubbing with recirculated water (or a solvent) for control of smaller particulate and to initiate the condensing process. It is at or slightly behind the Venturi that tars can begin to condense and can complicate the equipment design. Often, these Venturis have large liquid passageways and avoid areas where tars can accumulate. They are also designed with additional access points to facilitate cleaning.

The droplets formed in the Venturi are usually cyclonically separated using devices that expose the minimum of surface area onto which tars can accumulate. Vane or baffle type separators are often avoided.

The gases, now at or near saturation, are then subcooled to remove water vapor. The subcooling causes further condensation and the possibility of further tar build-up. The water vapor usually must be controlled, however, given the catalytic processes that often follow.

If the gasifier produces acidic gases (say, SO_2), an additional wet scrubber is used to absorb the SO_2 and neutralize it using an alkali (usually caustic).

All these devices may operate at high pressures (20 psig to more than 900 psig).

23.4 Primary Mechanisms Used

Cyclonic separation is commonly used to remove the dry particulate while the gas stream is at its hottest. Some systems reportedly use filtration. Venturi scrubbing (as described in previous chapters) is often used for fine particulate capture and to saturate the gas stream. In this application, the cooling also causes the condensation of tars.

Condensation is further applied to remove excess water vapor, since the goal is to produce syngas, not water vapor. In doing so, more tars can condense and some of the water-soluble hydrocarbons will be absorbed. The water from this stage is typically treated externally prior to being recycled.

These devices often run at a higher liquid-to-gas ratio than in more "mundane" applications.

Additives (both solid and liquid) may also be applied to sequester the tars. At this writing, various tar control techniques are being evaluated. In the meantime, the systems are designed for simplified maintenance access.

23.5 Design Basics

Gasification systems inherently deal with gas mixtures that exhibit low molecular weight and high specific heat. Also, if acid gases are evolved, upon dissolution a heat of solution/reaction occurs. To complicate things further, as pollutants and the carrier gas are cleaned and cooled, the gas properties change. Water of a molecular weight less than air, for example, is added and then is partially removed. Doing so changes the net characteristics of the gas mixture.

These gas streams are not like ambient gas streams. The molecular weight and specific heat must be calculated for the gas mixture at each stage. As mentioned in Chapter 1, "Air Pollution Control 101," psychrometric calculations are made for each stage. For these systems, the molecular weight and cp (specific heat) are most important. Table 23.1 shows a molecular weight and

TABLE 23.1

Molecular Weight and cp Estimate

Gas	Molecular Weight	Lbs/Hour	Molecules	Molecular Percent	Mass Ratio	Cp (Specific Heat)	Mixture Cp
H_2	2.02	1140.68	564.69	0.303	0.03436	3.468	0.1192
N_2	28.02	268.4	9.58	0.005	0.00808	0.2532	0.0020
CO	28.01	5971.81	213.20	0.114	0.17988	0.256	0.0460
CO_2	44.01	10,802.94	245.47	0.132	0.32540	0.248	0.0807
H_2S	34.08	0.68	0.02	0.000	0.00002	0.146	0.0000
C_2H_4	28.00	872.29	31.15	0.017	0.02627	0.083	0.0022
CH_4	16.04	2382.01	148.50	0.080	0.07175	0.520	0.0373
H_2O	18.02	11,742.32	651.80	0.350	0.35370	0.489	0.1730
NH_3	39.95	17.04	0.43	0.000	0.00051	0.125	0.0001
HC1	36.46	0.36	0.01	0.000	0.00001	0.137	0.0000
Total		33,198.5	1864.85	100.000			
Avg.	17.80	Dry lbs/hr = 21,456.21		Dry Mots = 1213.051		cp =	**0.4605**
Humidity Ratio:	0.547		lbs water vapor/lb dry gas		Dry MW =	17.68	

cp estimate for a gasification system gas stream for the gases as they enter the Venturi inlet.

This information is needed because many psychrometric programs allow you to adjust the cp and molecular weight, as these parameters determine the saturation conditions of the mixture. The saturation conditions in turn dictate the amount of water that needs to be added or removed, which determines the mechanical design of the system (pump sizes, heat exchanger duty, blowdown rates, etc.). Unlike many systems running with air mixtures, these calculations need to be performed for each stage (dry cyclone inlet; Venturi inlet; cyclonic separator inlet; condensing stage inlet and outlet; and if an acid control stage is used, its inlet and outlet).

If a compressor is used after the gas cleaning system, an accurate calculation of the gas properties entering the compressor is of critical importance since that device may be the prime mover of all gases through the system.

23.6 Operating Suggestions

The system components must be designed for simplified service access since the formation of tars is a given and periodic maintenance is often required. These systems can be and are designed, however, for extended campaigns before cleaning despite the challenges.

Instrumentation that monitors pressure drops through the system is used to forewarn of buildup issues. If a Venturi scrubber is used for particulate control, the Venturi is often designed to be free of any spray nozzles that may plug. In addition, Venturi designs that minimize the wetted surface (upon which tars may accumulate) are often used. Extra access doors and sometimes clean in place steam or solvent lances are built into the design. For the condensing stage, oversize packing is used, and designs that tend to produce a dripping liquid surface rather than films are generally favored. Some systems use spray towers or agitated fluidized bed type devices for the condensing sections to reduce plugging. Some systems have the provision for the introduction of solvents or oils to cut tar buildup in the condensing stage.

Specific requirements for operating the system are dictated by the type of furnish used. Some furnish produces greater quantities of tars than others; therefore, the designer must adjust a basic design to suit the realities imposed by that furnish. Current practice assumes that tars will indeed accumulate over time; therefore, the facility design includes maintenance shutdowns.

If the system has particularly problematic buildup areas (say, in ductwork), the area is designed to be easily removed. An example would be to flange smaller ductwork sections so that those sections can be removed for cleaning.

24

System Diagnostics and Testing

"The best laid plans o' mice an' men gang aft agley" (Robert Burns). Loose translation: Despite your best efforts, some things just go wrong.

This book focuses on selecting the proper air pollution control equipment for a given application. What happens if, however, you are "stuck" with what you have? Perhaps you do not need to replace it (or can delay replacing it). Maybe a repair or upgrade is in order.

Readers of the first edition of the *Air Pollution Control Equipment Selection Guide* requested that a chapter be added on problem solving for one of the most popular pieces of gas cleaning equipment, wet scrubbers. Given the popularity of that section, it was retained and updated based upon recent site-specific experiences. It is still true that the best of designs simply fail to work properly. Sometimes the process changes. Sometimes the economy changes and there simply are not enough funds to make a change. The pressure drop may be too high or the efficiency too low, chemical may be wasted, or assorted miscellaneous problems could arise. How does one go about correcting these problems? In this chapter, some proven ways of diagnosing and correcting some common and not-so-common problems with wet gas cleaning systems will be discussed. Many techniques mentioned also apply to dry scrubbers and may also be instructive for the reader.

24.1 Tools

First, you need the right tools.

To diagnose most scrubber problems, you could need any or all the following:

1. Copy of operating manual (or, lacking that, a copy of the purchase order and quotation for the equipment)

2. Copy of all stack tests on the system (particularly the most recent ones)

3. Copy of operating records, charts, datalogger output, etc.

4. Name and contact information of the person(s) operating the system

5. Name and contact information of the person(s) maintaining the system

6. A Dwyer Magnehelic™ gauge, or equal, of a range greater than the total differential pressure of the system (see Figure 24.1)

7. A pitot tube (such as the type S) and manometer (see Figures 24.2 and 24.3)

8. A pH probe and meter

9. An oxidation reduction potential (ORP) probe and meter if it is an odor control system

10. A 4–20 ma signal injector or calibrator if the system uses a 4–20 ma control loop.

11. A volt ohm meter (VOM)

12. Portable thermometer or temperature gauge

13. If available, a clamp-on liquid flow meter (rentable if you do not have one)

14. Tachometer (to check fan speed)

15. Clip-on ammeter (or another suitable meter to measure fan and pump amperage)

FIGURE 24.1
Dwyer Magnehelic for pressure measurement. (Dwyer Instruments, Inc.)

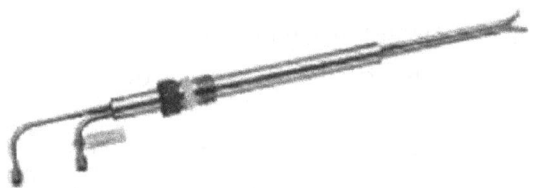

FIGURE 24.2
Type S pitot tube for gas volume measurement. (Dwyer Instruments, Inc.)

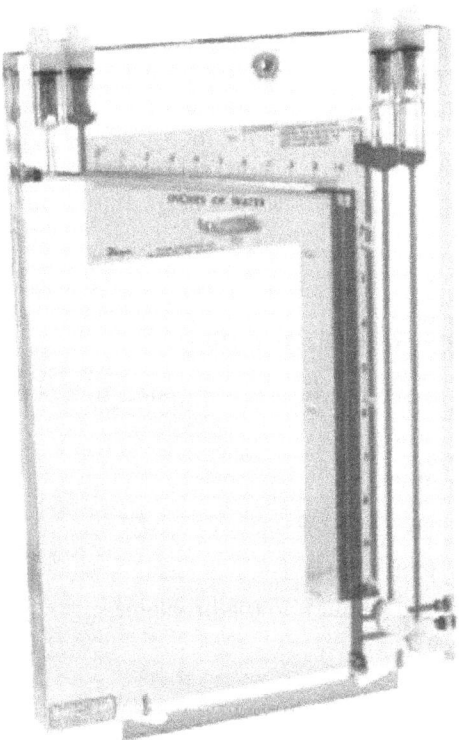

FIGURE 24.3
Manometer for velocity pressure measurement. (Dwyer Instruments, Inc.)

The Dwyer gauge allows you to measure the pressure loss across individual system components or, if the gauge has sufficient range, the pressure drop across the entire system. Changes in pressure can be indicative of mechanical difficulties (worn internal components, scaling, etc.) and are critical for your diagnosis. These gauges come in various ranges, and common ones for testing include a range of 0–2 inches water column (w.c.) and 0 to the maximum static pressure of the fan or other prime mover.

The one shown is a type S pitot tube. It is used for measuring the gas velocity pressure from which the gas flow rate can be calculated. A thermometer is also used to obtain the gas temperature. Given a measurement or an assumption of the gas humidity and the internal dimensions of the ductwork, the gas volume at the measuring point can be obtained. Accurate gas volume readings require at least four duct diameters ahead of the pitot tube and at least two duct diameters after the pitot tube. This is often not possible for diagnostics. Instead, a traverse of the duct is made (simply sliding the pitot tube from near duct wall to far duct wall) on the same plane and the peak velocity pressure is recorded. The average gas velocity pressure is usually about 80%–85% of that reading. If you are consistent in your measuring technique, the relative gas velocities through the system can be used to compare the gas flows at various points in the system.

The velocity pressure itself is measured using a liquid-filled inclined manometer as shown in Figure 24.3 or an electronic manometer. The readings are typically only in terms of less than 2 inches w.c., so a precise instrument is needed. In a pinch, a 0–2-inch Magnehelic may be used. The measuring device needs to be zeroed and in the case of the liquid-filled be leveled.

24.2 Isolate and Correct the Problem

A simple, multiple-step procedure is suggested to isolate and correct the problem:

1. Define the problem.
2. Inspect the scrubber and/or system.
3. Baseline the system.
4. Search for a *detailed* solution to the problem.
5. Fix the problem.

These steps are detailed in the remaining sections of this chapter.

24.2.1 Define the Problem

This obvious first step is often overlooked. Sometimes, we tend to run head-long toward solving problems rather than pause to think about what the problem *really* may be.

One way you can define the problem is by asking the right people the right questions. Who better to ask than the people running the system day after day?

Determine from the person responsible for operating the system the exact, if possible, behavior of the system that is unacceptable. Visible emission? High pressure drop? Excessive chemical consumption? Scaling? Interview both the operating person and the maintenance person (sometimes one and the same individual) to find out the type of problem at hand. Did the problem occur suddenly? Is it recurring? If so, at what intervals? Does it seem process related (triggered at a certain time of a process cycle)? Did it occur before and was temporarily fixed? If so, how was it fixed? How long did the "fix" last? What is the maintenance history? When was the last time any control instrumentation (usually pH and ORP meters) was calibrated? Any fan noises? Vibration? Odd sounds or other abnormal events (such as overheating, running pump dry, etc.)? Did the process change recently? Any major adjustments made recently? If so, what, and why?

After you get this information, find out what the system is *supposed* to do. Read the quotation and purchase order if available. Read the operating manual (*Note*: The number one cause of operating problems is the failure to read the operating manual). Did it *ever* do what it was supposed to do? If not, was the original vendor called in to fix it? If so, what was done? Who did it? You may have to call that person.

Sometimes, what the system is supposed to do is different from what the user expects or what the designer intended. This does not mean the system *cannot* do what might be expected, but it might mean that major changes are needed. Try to define the problem in the context of what the user expects.

24.2.2 Inspect the Scrubber and/or System

Make a visual inspection of the system. A "walk around" is the best method. Is what you see the same as described in the operating manual and/or quotation? If not, get an explanation from the people operating the system. If the unit or system was modified, find out why, when, how, and by whom. Get a phone number of the person(s) who made any changes (you might need to call them later to explain unusual changes in the system).

24.2.3 Baseline the System

To baseline the system, check all instrumentation and controls to make certain that they are within the intended operating parameters outlined in the quotation, purchase order, or operating manual. If possible, remove and calibrate all pH and ORP controls. Take a liquid sample and compare the readings to a calibrated pH and ORP meter (see Section 24.1, "Tools," above). If the readings differ, check the temperature compensation of the process probe (you may have to contact the vendor). Many modern pH controllers allow you to access the solution temperature from the controller's panel. Take a manual temperature reading using a thermometer or portable instrument. Do they agree? Many times, a false pH reading is caused by a failure of the temperature compensation circuit. If the readings do not agree, check the probe-to-controller wiring. Often a loose wire from the temperature compensation circuit can cause a false reading.

If the system has a recycle flow and/or blowdown flow indicator, does this agree with the design? A clamp-on Doppler type flow meter can be a handy check of these flows. If the flow rate is out of specification, note if the flow is higher or lower than normal. If lower, the problem could be a simple valve setting, line plugging, or pump impeller problem. If higher than normal, a valve may be too far open, or some flow resistance (such as a spray nozzle or two) that had been present is now gone.

Try to set all operating parameters as near to the design as possible. Many times, after doing so, the problem will resolve itself. If the problem is still present, you need to get into the details.

24.2.4 Search for *Detailed* Solution to the Problem

You should have enough background information at this point that you can focus on the details and isolate the problem. In doing so, you will likely find that you have multiple small problems, effectively in layers. Go after the most obvious ones first. Sometimes, fixing the obvious cures the hidden problem as well.

Inspecting certain subsystems can isolate common problems. These subsystems include the following:

1. The gas cleaning device itself (the inlet flange to outlet flange unit)
2. The liquid circuit (pump, valves, piping, nozzles, etc.)
3. The instrumentation (pH and ORP controls, level controls, etc.)
4. The gas moving device (usually a fan)
5. The gas discharge device (usually a stack)

24.2.4.1 Gas Cleaning Device

It would take many additional chapters of this book to describe various minor problems with wet scrubbers that could cause improper performance. Some common problems are addressed in this section. If you compare the suggestions to the symptoms you find during your Interview and Inspection, you may find the solution to at least part, if not all, of your problem. With certain exceptions (such as mist chamber scrubbers), most wet scrubbers exhibit a gas pressure loss through the unit. The loss of pressure is the result of changes in direction or acceleration or deceleration of the gas in the scrubber.

24.2.4.1.1 Lower than Normal Pressure Drop

If the pressure loss is lower than design and the gas and liquid flow conditions are at or near design, this indicates a change in an internal device. For example, a baffle that once was in place may now be corroded or eroded away. In cyclonic separators, the loss of an inlet baffle could be a contributor to an entrainment problem. An internal inspection of the scrubber would reveal the condition of such a baffle. For packed towers, a lower-than-normal pressure drop could indicate gas channeling in the packing caused by improper liquid distribution. An improper or insufficient spray of scrubbing liquid into the unit could cause a lower-than-normal pressure drop in a dynamic scrubber (such as a sprayed fan design).

If the gas flow is inherently lower than the original design, contact the original vendor or a scrubber consultant to have them estimate the expected pressure drop at that new flow condition. Most scrubbers have pressure drops that are in proportion to the gas flow rate. If you cannot find someone to estimate the expected pressure drop, factor the pressure drop by the ratio of the current gas volume to the design gas volume. If the result is a pressure drop that is still too low, this indicates a problem with an internal component that alters the direction of the gas or is intended to accelerate it (such as a Venturi throat, orifice plate, impingement plate, etc.). Focus your attention on that type of device.

An example is a tray type scrubber that exhibits a lower-than-normal pressure drop after having operated close to design for years. Many of these tray towers have removable trays consisting of small, perforated openings (often less than 3/16-inch diameter). If these trays lift or shift (say, from a sudden puff of gas or other upset), they may not return to a seated and sealed condition. The pressure drop will be abnormally low. The remedy is to remove and straighten (or even replace) the trays, making certain that they are fully seated.

A low pressure drop in a tray scrubber that uses internal liquid management weirs or downcomers (seal boxes) could be caused by a failure of those devices. Weirs are sometimes used at the end of a tray to hold up a minimum depth of liquid. If the weir corrodes, the liquid depth will decrease, yielding a lower pressure drop. If a downcomer (which lets liquid descend from an upper tray but normally prevents gas from bypassing the tray) loses seal, the pressure drop will be lower. The remedy is to increase the seal leg depth. This often requires an internal inspection. Some firms offer X-ray or ultrasonic inspections of such towers "on the run," thus allowing such problems to be more easily diagnosed.

A sprayed fan scrubber might exhibit a lower pressure drop if the inlet spray nozzle pattern changes or the nozzle plugs. Usually, the gas volume will increase along, yet the fan amperage may drop (since the fan is not moving the water as before). Using a suitable ammeter, check the fan amperage versus design. Check the liquid flow (see next section). You could find that you have a nozzle with an altered spray pattern (say, from scaling). Replacing or cleaning the nozzle may restore the scrubber to its design performance. Usually, the sound of these scrubbers will change if the wheel is improperly loaded with liquid.

A packed tower exhibiting gas channeling usually results in poor removal efficiency since the gas bypasses through the scrubber essentially untouched by the liquid. If you open the top inspection door (if provided) or can otherwise see the top of the packing, you may see that it is displaced (you will see a mountain or a valley of packing—it will not be level). High gas velocities can throw packing from the center of the tower toward the wall, leaving a center valley. Excessive liquid pressure in towers equipped with spray type liquid distributors can blast the packing toward the center, leaving a mountain of packing there. The solution to the problem may involve improved liquid distribution and/or lower liquid pressures or the addition of a hold-down grating to hold the packing in place. If the tower has more than 8 feet of packing, the problem may be caused by the lack of a redistribution to move liquid from the wall back to the center of the tower.

On Venturi scrubbers, wear, or damage to a component in the throat zone (restriction) usually causes low pressure drop. Often, a simple plate is used (see Chapter 19, "Venturi Scrubbers"), and it wears away over time. Replacing this element often solves the pressure-drop problem. If the throat is intact and at or near specification, the problem usually is liquid-distribution related. For throats greater than 12 inches in width, a center spray is often needed to fill the center area of the throat with liquid.

For cyclonic separators, an incorrectly sized tangential inlet baffle is usually the cause of low pressure drop. These baffles also wear away with time. Another symptom is entrainment from the separator. Often, on induced draft systems, the drain leg from the separator may not be sealed under liquid. This can allow air to be drawn up through the drain, reducing the overall pressure drop of the unit and usually causing erratic pressure drops (sometimes, even noise from the vessel). Sealing the drain under liquid usually solves the problem.

24.2.4.1.2 *High Pressure Drop*

Higher than normal or expected pressure drop usually means that a restriction has been created, perhaps suddenly or over time, in the scrubber. A plate could shift and restrict a gas passage. Scaling from water hardness or other chemical reactions inside the scrubber could cause gas flow restricting effects that raise pressure drop.

On Venturi scrubbers, if the throat is adjustable, the movable portion may have been set too far closed. Obviously, the solution is to open the throat until a suitable pressure drop is obtained.

On cyclonic separators, if the sump level rises to the level of the gas inlet, the liquid could offer additional gas flow resistance, resulting in excessive pressure drop. Lowering the liquid and/or installing a sealed overflow connection to prevent recurrence of the problem usually is a solution. Other times, the swirling of the liquid in the separator sump can cause the liquid to rise at the wall, restricting airflow and usually causing entrainment. Antiswirl baffles under the liquid level are a typical solution to that problem. These baffles are normally 6–8 inches high and run from the periphery of the separator sump toward the drain opening such that any swirling liquid must jump over this baffle. Most often, only one or two such baffles are needed.

Packed towers usually exhibit high pressure drop at design gas and liquid flow rates if the packing is plugged. Some systems can be acid or caustic washed (sometimes on the run if the operating permit allows such maintenance) by adjusting the recycle pH temporarily. Others must be shut down entirely, the packing removed, and replaced with new or cleaned material. Investing in a water softener for the make-up water can sometimes reduce chronic scaling. This is common in hydrogen fluoride scrubbers where a carbonate and silicate reaction can produce troublesome hard scale.

Dynamic scrubbers showing high pressure drop usually have restrictions in the gas outlets from the devices or exhibit higher than normal liquid rates. The motor amperage usually is also high if the liquid is to blame.

Tray towers have high pressure drops if the tray openings begin to plug. Sometimes the trays can be removed and be cleaned. Often, however, the tray pluggage requires replacement of the trays with new ones.

24.2.4.2 Liquid Circuit Problems

Pumps are the most common liquid "movers" in wet scrubbing systems. Since the liquid itself and how it is administered are the greatest contributors to the performance of the scrubber, the liquid circuit must get special attention. Experience has shown that pumps are reliable devices when they receive proper maintenance. The pump supplier's maintenance procedures should be followed to obtain the best performance from the pump(s). How the liquid is administered into the scrubber, however, varies by design. Therein lie many problems. If the scrubber removal efficiency is a problem, the first point to look at is where the liquid meets the gas, then work backward to the pump. Most problems can be localized to improper liquid rate or distribution at the point of injection.

For example, in spray tower scrubbers, a primary cause of lower scrubber efficiency is nozzle plugging or failure. If a nozzle plugs or breaks, the spray pattern is no longer of the type or shape required by the designer. This change may adversely affect performance. Sudden pressure drop or flow changes as revealed by a system datalogger may indicate the mechanical loss of one or more nozzles. A gradual decline in flow or increase in pressure drop could indicate a slow plugging of the nozzle or its supply header. Often, the tower must be shut down and the nozzles be individually inspected. Spare nozzles are a good insurance policy for spray scrubbers.

In packed towers with spray type liquid distributors, the same holds true. A more common problem, however, is an insufficient number of nozzles, that is, poor liquid distribution. You may see this if you observe the top of the packed bed and see spray-pattern-induced discoloration or packing solids buildup. The remedy is to measure the existing spray pattern and design a new one with fully overlapping coverage. Sometimes, wider spray-angle pattern nozzles are all that are needed.

A more common occurrence in spray distributor type packed towers is that too much packing has been installed in the tower. The distance between the packing and the spray nozzle is too low to allow the spray pattern to form. Another problem is the use of nozzles (so-called pig-tail type) that produce a rose petal pattern near the nozzle and a more uniform pattern away from the nozzle. These nozzles work well at the intended nozzle-to-packing distance but can be a problem if that distance is reduced. If the packing must be up close (say, to get greater removal efficiency), check with the nozzle vendor for a replacement selection that produces a full cone pattern in the nozzle or in proximity to the nozzle discharge.

On Venturi scrubbers, the liquid headers are usually open pipes and plugging is the only concern. On annular types that use a center pipe, the pipe must be centered on the apex of the cone. Typically, one third of the liquid flow goes to the center pipe and one third goes to the tangential or wall headers.

On fluidized bed or ebulating bed scrubbers, the headers are typically horizontal with open holes drilled horizontally through both walls. These headers are usually "dead-ended", and solids can build up in the far holes. These headers are removed and cleaned to correct the problem. Care must be taken to replace the headers so that the liquid discharges horizontally, not vertically, on most designs.

Flow control valves (especially bleed control valves that open and close) are a common source of hard-to-diagnose problems. They usually create an intermittent or randomly occurring problem. These can sometimes be taken out of the control loop and a 4–20 ma signal injected from a suitable instrument to "fool" the valve into operation. Usually, the positioner for such a valve will not permit full travel and the valve either does not open or does not close fully. Limit switches are commonly used to reverse the action on these devices, and a bit of electronic troubleshooting using a VOM can isolate the problem. Such testing must, of course, be conducted in compliance with the safety and operating constraints of the system. Most often, such tests are conducted when the system is down.

Scaling of recirculation piping is one of the most common problems. If chemical washing is not effective, consider changing the piping design to allow simplified removal of problem-prone sections. Some scrubber chemical suppliers offer additives that can sequester the type of scale-forming solids and allow them to be flushed away. These companies, to date, include Nalco Chemical, Betz Dearborn, Quaker Chemical, and many others. The use of these chemicals must be compatible with your water treatment method or sewer regulations.

24.2.4.3 Instrumentation Issues

Modern-day instrumentation includes built-in diagnostics and troubleshooting methods. Many plants have in-house instrumentation personnel who add their considerable experience to that provided by the vendor. It has been found that most instrumentation problems that occur on a previously properly operating system involve the simple opening of circuits. This could be caused by loose or corroded wiring or by the failure of circuit boards in the respective controllers.

For the former, the VOM and 4–20 ma loop calibrator are useful tools in diagnosing problems. The VOM can help you find open circuits, and the loop calibrator can permit the injection of a known signal so you can evaluate the response.

As mentioned earlier, pH probes and controllers typically use a temperature compensation circuit that sometimes opens. Usually, a wiring connection opens, but sometimes it is in the probe itself. Not only should the probe be calibrated but its measured temperature should be confirmed. This function is usually a menu item on the controller panel. On older systems, you must check for an open circuit in the temperature compensation portion of the wiring. In five-wire probes, these wires are usually delineated in the operating manual and can be traced easily.

The loop calibrator can be used to check operation after the device is separated from the circuit. Calibrators vary, so one must check the documentation associated with the device. A common problem can be revealed when one investigates any recent instrumentation changes that may have been made. Sometimes, a ground loop is accidentally formed, which provides a false reference voltage for one or more controllers. This is quite common with level control systems that are immersed in conducting solutions (such as resonant frequency or capacitance types). The remedy sometimes involves the installation of a grounded reference electrode in the sump.

Instrumentation vendors want their equipment to work properly. Most have toll-free technical assistance numbers. It is a good idea to list these numbers along with the model and serial number of the system's instrumentation in the front of any operating manual. It is also suggested to have spare controllers or at least output cards for the key system control units. Often, local welding or voltage spikes can accidentally destroy output cards. Isolated output buses are also available from instrument vendors to fix chronic problems regarding spurious signals.

If your system uses draft control via a draft sensor (load cell, etc.), the loop calibrator can be used to set this device up prior to startup. Often, the control does not range from 4 to 20 ma. Usually, zero draft is 12 ma. You need to contact the controller vendor to determine the set-up sequence. The common method is to calibrate the controller for the maximum draft (negative reading) at 20 ma and enter a positive pressure at 4 ma. The SPAN of the controller is used to provide smooth transition through zero pressure.

Many controllers sample (i.e., read a signal) at a rate different from their output control. This is sometimes called RPM, or resets per minute. Basically, a reading from the sensor is followed by an output signal to the controlled device (say, a flow control valve, draft control damper, etc.). A controller that behaves erratically with its output signal resulting in control element hunting or overheating can often be dampened by decreasing the RPM set point at the controller. The controller simply waits longer between sampling sessions. Most scrubbers do not need constant correction. Once every 20–30 s is usually sufficient. Exceptions would be processes that exhibit sudden puffs or surges. In those cases, more frequent resets would be suggested.

Draft instrumentation is typically accommodated using Dwyer Magnehelic gauges or similar devices. These Bourdon tube type or diaphragm type devices can fill with condensate from the scrubber since most scrubbers saturate the gas stream. If this occurs, the gauge will read both the actual pressure and the head of liquid in the sample line. Sample lines should come out of the pressure tap, then up much like a P trap seal, then down below the gauge. A drain tee and valve at the low point usually allow the conden-sate to be drained. The sample tube would then go back up to the gauge. These problems are very evident on systems used in cold climates. Periodic draining of the line is required.

If compressed air is used to blow back the sample line, do it after discon-necting the gauge; otherwise, you could damage or even destroy the internal mechanism of the gauge.

Keeping a record of your instrumentation service events can be very handy in determining where your maintenance dollars are being spent. Investing in an ultrasonic probe-cleaning device, for example, may mitigate cleaning pH probes daily. Continual scaling of level controls could lead you to changing to a noncontact type control. The service record will help reveal these pos-sible money-saving opportunities for you.

24.2.4.4 Gas Moving Device

Usually, centrifugal fans are used to move gases to and through a wet scrub-ber. Exceptions are the use of eductors, wherein the motive power is pro-vided by the pump pressure and the momentum of the liquid through the eductor. Some other systems exhibit a process pressure high enough to move gases through the scrubber.

24.2.4.4.1 Fans

Industrial centrifugal fans are typically of radial blade designs where sol-ids are present and backward inclined or radial tip designs where only gas is treated. Fiberglass-reinforced plastic fans sometimes have special wheels, given the centrifugal reaction forces on the wheel.

Fan problems are usually related to balancing (vibration) and alignment. Fans that constantly go out of balance are usually being impacted by par-ticulate. The source of the particulate must be controlled, or if that is not pos-sible, the fan wheel type must be better matched to the application. Often, a periodic fan wash using an inlet nozzle is required. The discharge stack or duct, however, must be designed to accommodate the spray that is generated. With the advent of field laser alignment technology, fans can be set up within exacting alignment tolerances. Laser alignment once per year as a minimum is suggested for any fan developing more than 30 inches w.c. static pressure.

Some fans have an adjustment for the gap between the inlet bell (gas inlet) and the wheel. This is sometimes called the "wheel face to cheek" dimension. On high static pressure fans, as the fan runs up to speed, the fan inlet can be drawn slightly into the fan wheel. The gap allows this movement (along with thermal expansion). If the gap is too large, however, the wheel can start to "windmill" and not develop the proper static pressure. With wind milling, some of the gas moved by the fan simply short-circuits to the gas inlet. This gap should be measured upon installation and reset every time the wheel is cleaned or removed.

On belt-driven fans, the inboard bearing usually is a tapered roller bearing or otherwise has a thrust bearing. The outboard bearing is typically a floating bearing to allow thermal expansion. Too often, when the bearings are changed, the order is reversed. A floating inboard bearing allows the wheel to move toward the fan inlet and can sometimes cause contact as the bearing wears. The result is noise and possibly wheel failure. The bearings must be professionally installed.

A more common problem, at least at startup, is incorrect fan rotation. This can occur during a change of the motor control center or starter. A fan can move air even if it is rotating backward. Changing one of the fan leads corrects the rotation on three-phase motors.

The fan vendor's literature, though usually quite sparse in content, should be followed for fan servicing. Particular attention should be paid to the lubrication program.

Fans usually have housing drains, particularly when the fan is applied to a wet environment. A common occurrence is that the plug for the drain is missing, thus allowing outside air to enter. If the fan is handling a wet gas stream, the fan drain should be piped to a sealed drain, that is, through a trap so air cannot enter. Depending upon its location on the fan housing and downstream gas flow resistance, the drain may be under positive or negative pressure versus atmosphere. The trap can be quite high on many systems. One way to check is to make a threaded replacement plug with a tap to a pressure gauge. Install the plug and read the pressure when the fan is moving cold air (cold static) if possible. The seal leg should be at least 10% higher than this reading in inches of water. If you cannot get a reading, use the fan performance curves, and measure the outlet static pressure. The drain is usually at least at the outlet static pressure.

24.2.4.4.2 *Eductors*

The key element of an eductor is the nozzle. If the nozzle does not produce the proper spray pattern or maintain the correct flow rate, the draft produced, and the flow produced can be insufficient. Causes of reduced gas flow rate usually are the result of low liquid rates to the eductor. Reductions in draft are typically caused by poor spray patterns from the nozzle, that is, a partially plugged nozzle. In either event, the nozzle should be inspected and cleaned or replaced if low flows or low drafts are experienced.

Another potential problem is wet/dry line buildup just ahead of the restriction in the eductor (throat). The wet/dry line buildup occurs at the point where the spray meets the converging section of the eductor or at a point just inside the beginning of the throat. If the gas stream contains some particulate or the liquid contains dissolved or suspended solids, the solids can migrate to the wall of the scrubber and stick. This layer can build up to the point where a gas flow restriction is created, reducing the gas flow through the eductor. The remedy is to clean the converging section. If this problem is chronic, the converging or "approach" section can be made fully wetted by adding tangential liquid inlets that swirl the liquid much like the action in a dentist's bowl. The added liquid, however, may reduce the momentum effect of the spray liquid. It is best to contact the eductor vendor regarding any modifications.

Throttling the exhaust can also reduce the flow through an eductor. If excessive backpressure is created by downstream equipment, it can reduce the flow through the eductor. Sometimes increasing the liquid flow rate through the eductor can compensate for downstream resistance. Pressure losses downstream of 4–6 inches w.c. should be overcome by most eductors. If the gas flow is low but the eductor is working well, look at the equipment or ductwork located downstream of the eductor to see if excessive restrictions (such as an overfull tank or closed damper) could have occurred.

24.2.4.5 Gas Discharge Device

Exhaust stacks are the most common discharge devices. They look so simple. They can be a major problem source if they are not intelligently designed. With a wet scrubber, the gases are saturated or should be nearly saturated with water vapor as they leave the stack. Exceptions are those cases where an exceedingly high static fan is used, or a gas reheat system precedes the stack. The heat of compression of a high static pressure fan can heat the stack gases about 4°F for every 10 inches w.c. of fan static pressure. Heated gas can hold more moisture than the colder gas. Though the gas enters the fan saturated, it can leave below saturation.

Since the gases are near saturation, any cooling of the gases can cause condensation. If the stack gas velocity is excessive, the condensation that is formed can go up the stack and be carried out. Worse yet, these droplets can accumulate particulate and be caught in the sampling train only to be counted as gas-borne particulate.

As a result, many wet scrubber stacks are limited to vertical gas velocities of less than 40 ft/s. Experience has shown that condensation entrainment can rise in a stack at or above this velocity. Stack velocities of 35 ft/s are normal design figures.

If a stack test is failed by high front-half particulate and the testing firm notes excessive probe wash and water accumulation, you could have a stack problem. Check the stack velocity. If it is over 40 ft/s, you likely have an ascending wall entrainment type problem. This can be determined by making a small test probe from a piece of pipe (see Figure 24.4). The slot in the probe is oriented so that water rising in the stack near the wall is diverted into the slot of the probe and drains out (watch out, this water is likely to be hot on systems attached to high-temperature sources). If the probe is turned so the slot faces upward, it will catch water descending the stack wall. Descending wall liquid is usually normal and does not cause a problem unless the stack lacks a drain or has some internal obstructions that the liquid can hit and be splashed into the gas stream.

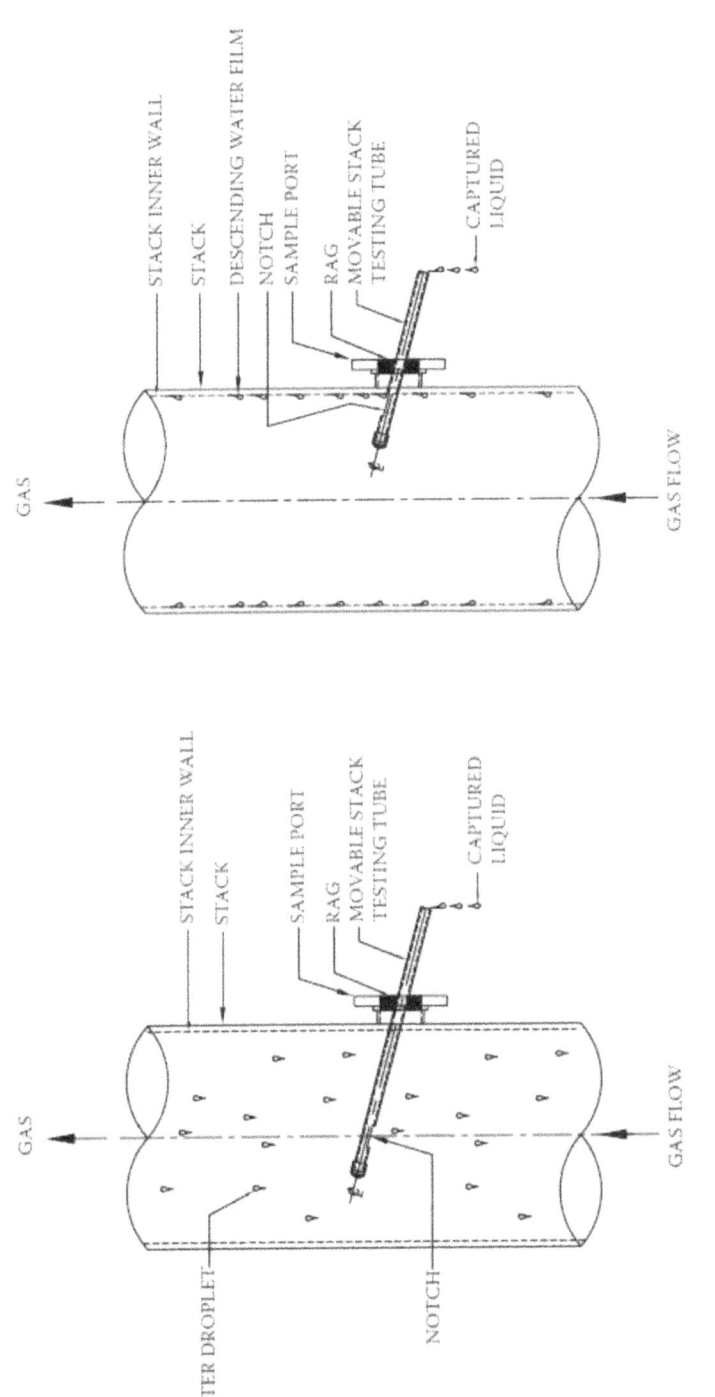

FIGURE 24.4
Method for checking stack entrainment.

If the probe, slot up, is pushed into the stack farther (away from the wall), few if any droplets should drain out. If the probe slot is turned so that it faces downward, it should catch liquid spraying upward if the gas velocity is too high or there is entrainment from some upstream device. This is called ascending center entrainment.

If you detect ascending center entrainment, you likely have more than one problem. If you also measured ascending *wall* entrainment, your stack is likely too small, *and* you have some sort of upstream entrainment problem. If you did not measure ascending wall entrainment, the stack is likely to be of sufficiently low velocity, but you have another entrainment producer.

A common error is to discharge the fan exhaust directly into the stack base. If the high velocity fan exhaust hits the condensate that drains down the stack, a spray could result that sends the condensate up and out of the stack as entrainment. The fan discharge, sometimes called an *evase* (E-vah-say), should extend into the stack and be equipped with a brim (that looks like a horse collar) that moves the descending liquid around the inlet. The stack should also have a suitably sealed drain. The drain should be sized to handle about 3%–5% of the recycle liquid of the scrubber if possible. This would allow recovery in case a gross entrainment accident occurs.

24.2.5 Fix the Problem

Fixing the problem can be inexpensive and quick, expensive, and slow, or any combination between. With most scrubbers, the actual work entails generally available trade skills such as piping, electrical work, mechanical adjustments, and so forth. If you make a task list or action item list, you can see the extent of the corrections needed. Try to organize them by manpower requirements and expense. A welcome suggestion is to fix the most easily corrected problem first and then evaluate the results. This is obvious but doing so will not always solve the problem. Many times, it does produce the desired results.

If your task list is long, you should not avoid confronting a decision to replace the scrubber. It could be that the system was never matched properly to your real problem and will never meet the desired requirements. The unit may have been the least expensive that was available at the time, the process could have changed, the water quality could have deteriorated, or the system could have been overwhelmed and is simply too small. It could be plain worn out. Replacement could be the best decision.

As mentioned at the outset of this chapter, there are many possible problem areas that need to be explored in diagnosing problems regarding wet scrubbers and the ancillary equipment required to run them. Though each problem is quite simple, the sum of them can sometimes be quite intimidating. It is hoped that some of the preceding suggestions will help you localize and solve problems you may be having any may allow continued operation of the air pollution control equipment you already have.

25

Optimization of Gas Cleaning
Equipment, Overview

25.1 Defining the Problem

Let us say you recently installed some wet scrubbing hardware. Now you are focusing on getting that equipment to function as best as possible with your application. Or maybe you have some old equipment that you want to "tweak" to improve its performance and perhaps extend its useful life.

Once equipment is selected, purchased, and installed success often hinges on the optimization of that equipment. Sometimes, the equipment selected does not easily "fit" the process or the process inherently changes its emission characteristics during its operation. In the latter, the air pollution control device, though selected for one condition (say, the maximum emission rate) may not be best suited for other, though "normal" operating conditions. Conversely, a device sized for "normal" conditions may exhibit reduced performance under peak or upset conditions. A sharp reduction in gas volume, for example, may adversely affect the performance of a device that relies upon gas velocity to achieve the desired performance (such as a dry cyclone). Similarly, the droplet separation efficiency of a cyclonic separator or chevron type droplet eliminator may be reduced under certain conditions.

25.2 Post Installation Optimization by Equipment Type

Thus, post-installation, air pollution control equipment often needs to be optimized. By "optimized", in the context of air pollution control, we mean that all things that influence the desired performance of the equipment are *maximized* and all things that detract are *minimized*. By "maximized" and "minimized" we mean that there will likely be some design or operational parameter that detracts from achieving total 100% performance. Reality adds chaos to most any separation technique. Particles do not always follow expected trajectories. With some particles, during their separation from a gas stream, their contact and interaction can cause size reduction making those smaller particles harder to remove. Gases do not always behave as expected. Liquid does not always distribute and move as either theory or practice predict.

Bounded by the laws of physics, the application of hardware is sometimes like "fitting a square peg into a round hole". Some "sanding and fitting" is often required. In addition, control of the application to which the hardware is applied is really the responsibility of the user. The hardware vendor basically designs and applies that hardware to suit the user's description of the application. The project's success is therefore defined by the proper coordination of the process parameters with the air pollution control hardware. Candid communication between the user and the hardware supplier is thus required.

The reality is that we do not know everything and what we do know is subject to revision as we learn. The optimization of the air pollution control hardware can lead to the success that both the user and the hardware supplier desire. The following chapters provide suggestions as to how to gain the best performance from the hardware you have selected.

26

Dry Cyclone Optimization

26.1 The Dry Cyclone

Of all the "hardware" available to be applied to air pollution control, the dry cyclone often handles the most particulate and performs that task in the simplest of manner. Dry cyclone collectors are often the first device used in a multi-device gas cleaning train. This means the dry cyclone sees the highest *concentration* and most often, the *highest level of erosion and abrasion*. As mentioned in the Dry Cyclone chapter (Chapter 4), the target dust that the dry cyclone must separate from a gas stream may also be "friable" (reduced in size as the dust wears on itself and internal cyclone surfaces). Relying on the fact that solid particulate wants to move along a trajectory (unless that path is altered), the mechanical design of a dry cyclone is intended to move that path out of the carrying gas stream for collection. In effect, the particles want to move straight ahead, and the cyclone says "no", you must move in a curved path. Centrifugal force is applied to alter the direction of the particulate towards the vessel wall wherein the particulate moves out of the gas stream. Simple.

A dry cyclones installation is shown in Figure 26.1 (Lundberg). The arrangement shows four (4) cyclones mounted in parallel. The inlet ductwork is to the left of the center of the photo. The discharge ductwork is at the top.

26.2 Dry Clone Performance Optimization

Some things can be explored however to optimize the separation efficiency of a dry cyclone.

Some potential problems are as follows:

1. Nonuniform flow of particulate into the cyclone gas inlet caused by upstream disturbances.
2. Particulate size reduction resulting in increased separation difficulties caused by excessive local stream velocities.
3. Lack of true vertical vessel orientation creating an unbalanced gas flow.

FIGURE 26.1
Dry cyclone. Lundberg.

4. Rough or nonuniform interior surfaces in the vessel.

5. Excessive dust hopper level.

6. Existence of dust buildup points at flanges and other fittings resulting in reduced dust separation.

7. Air infiltration through particulate (dust) discharge devices.

8. Static electricity buildup.

26.3 Importance of Uniform Gas Flow

Optimization efforts to explore may include:

> To provide a uniform flow of particulate and carrying gas, the ductwork leading to the cyclone gas inlet should be a straight as possible. Though the installation location may be the final arbiter, if possible, at least two straight duct diameters should be present ahead of the cyclone gas inlet. If the main ductwork is round and the dry cyclone gas inlet is rectangular, any transition should be achieved well before the cyclone gas inlet. Some installations use an eccentric type transition wherein the lower portion of the transition is tangent to the lower portion of the dry cyclone gas inlet. The latter is applied since the particulate tends to flow by gravity more towards the bottom of ductwork than higher up. By using an eccentric transition, the particulate can flow uniformly into the dry cyclone for separation rather than being "kicked" up in the transition.

26.4 Minimizing Particle Size Reduction

> For some particulate, movement of the particulate in proximity with other particulate can cause a particle size reduction. As the particulate is size reduced, the removal efficiency decreases. Most dry cyclone

suppliers will have taken the "friability" (size reduction tendency) of the particulate into account during the sizing of the dry cyclone. Those vendors use empirical data and other experience to alter the geometry of the dry cyclone(s) they provide. During operation, however, conditions can change that affect the particle size. To optimize the cyclone performance, a particle size distribution and gas velocity reading should be taken, and that data is provided to the cyclone supplier. Upon a review of the data, the supplier may recommend changes to the dry cyclone gas inlet. For example, instead of a 90-degree gas inlet, an involute type gas inlet may be suggested. The involute may wrap around the vessel 180 degrees or more thus injecting the gas stream in a less abrupt manner thereby reducing size. They may alternately find that an additional cyclone stage may be needed.

26.5 Importance of Vertical Orientation

Dry cyclones in vertical orientation should be truly mounted vertically. This somewhat obvious fact is often overlooked or given settling of the support structure, change over time. The Operation and Maintenance manuals for dry cyclones highlight the need for true vertical orientation. If the cyclone is not truly mounted vertically, the cyclonic gas flow path can be upset. The ascending vortex in a dry cyclone inherently tends to form around a vertical "virtual" axis. If the vessel is not truly vertical, the spinning gases can impinge randomly at the vessel wall thereby upsetting the separation of the particulate from the gas stream. To adjust the position of the vessel shims may be needed at the support lugs or flanges to achieve true vertical orientation. If the application involves hot gases and expansion and contraction produce forces that move the dry cyclone away from vertical, expansion joints may be required in the gas inlet and outlet ductwork. Some installations using hot cyclones hang the dry cyclone vessel from spring-mounted hangers to allow for the expected movement.

26.6 Smooth Interior Surfaces

Usually, the supplier makes sure that the interior surfaces of the cyclone are sufficiently smooth for the application. The surface characteristics may range from simple post weld clean up to a polished surface. It is at the interface of the cyclonic gas stream and the vessel wall that inefficiencies can occur. If a dry cyclone vessel has been serviced or repaired, to optimize the performance the interior surfaces should (if those surfaces are accessible) be brought back to the original level of quality or better. Any appurtenance or rough area should be repaired. Any dents that could impair the smooth motion of the gases should be repaired.

26.7 Maintain Control of Dust Level

As the dust level in the dry cyclone dust disengaging hopper (if so equipped) increases, the ascending vortex of the dry cyclone discharge tube can come near the dust. Though the gas stream is spinning, this vortex has a vertical gas velocity component. The vertical component tends to lift the dust back up the gas discharge tube thus reducing the separation efficiency. Typically, the dry cyclone designer allows for an enough gap between the expected highest dust level and the lowest point of the ascending vortex. If the dust level becomes excessive however some of the collected dust can be entrained up and out. To minimize this from occurring, the dust level in any discharge hopper should be carefully regulated. This task can be accomplished in several ways. If the hopper is equipped with a dump type discharge valve (or valves in series) adjustments can often be made to the counterbalances based upon operational experience so that the dust is removed uniformly. If the hope is equipped with a rotary lock, the lock's speed may need to be adjusted. To monitor the hopper dust level, not contact type dust bin monitoring instrumentation may need to be added.

26.8 Trim Gaskets

An often-overlooked area that can prevent the dry cyclone from operating in an optimal manner is the effect of the gaskets on particulate build up and discharge. The gasket material should not, in any location, extend beyond the interior vessel wall. If the gasket material is exposed, not only can dust be built up upon the gasket surface and be discharged, but the uniform gas/particulate flow at that point can be disrupted. Some firms optimize the dry cyclone operation by removing any full-face gaskets and replacing the gaskets with rope type wherein the rope is applied within the bolt circle of the flanges. Doing so eliminates the possibility that gasket material may impede the separation efficiency of the dry cyclone. This technique can also be applied to upstream and downstream ductwork.

26.9 Minimize Dust Discharge Device Leaks

If a dry cyclone suddenly exhibits a reduction in removal efficiency, the primary cause is likely a leak at the dust discharge device. Air infiltration at this point (assuming the dry cyclone is operated under negative pressure) can cause a drastic reduction in removal efficiency. To check for leakage, if the dust discharge point is accessible a simple smoke test will reveal any inward motion of ambient gases into the dust discharge device. The vertical velocity component of the ascending vortex in the dry cyclone is now supplemented by the vertical velocity of the air stream leaking passed the dust discharge device. If dump or "double dump) dust discharge devices are used, wear can cause leakage. In addition, the dust itself is used to provide a "seal". If the dust hopper has insufficient dust and insufficient seal can result. Sometimes holding the dust discharge valve closed for a period and reducing the counterbalance force and restore the required inventory of dust that provides the seal (refer to the supplier's literature regarding adjusting the dust discharge valve(s)). If the hopper is equipped with a rotary lock, the sliding seals in the lock may need to be replaced or be adjusted. If the lock is variable speed, the rotational speed may require adjustment. Some locks run at excessive speed which not only prematurely wears out the seals but also does not retain enough dust between the lock's internal cavities to affect the required seal. To optimize the dry cyclone operation there should be no inward leakage of air.

26.10 Control Static Electricity if Present

A more subtle effect that can reduce dry cyclone performance is static electricity buildup. This so called "piezoelectric effect" is caused by particulate passing over adjacent particulate causing an electric charge to be produced. Not all particulate exhibits this characteristic. Dry cyclone vendors can compensate in their design but can only do so is an accurate description of the particulate is provided during the design or optimization. Sometimes bench scale testing is required. If this charge causes the separation distance between particles to increase, reduced separation efficiency may occur. There exist various techniques to remove this charge but he most basic includes directly ground the dry cyclone vessel. By "directly grounding" it is meant that the vessel is configured with its own, dedicated (not shared) ground. Additional lugs for grounding may be applied to the vessel. Sometimes a grounding probe is added to the dust hopper to remove the static charge. By removing the static charge, the dry cyclone's performance can be optimized.

26.11 Adjust for Particle Size Differences

If the particulate is "friable" different dust size characteristics may occur within in the same dry cyclone. The gas inlet, for example, may be populated by large particulate while the zone in the hopper or cyclone discharge may be a mixture of both the captured large particulate plus the smaller, size-reduced particles. To optimize the cyclone performance, the engineering optimization thought process must recognize these size differences and adjust accordingly. The dry cyclone supplier likely has suggestions for optimization based upon size for their specific design.

The following is intended to expand upon and supplement the suggestions that were explored in the Dry Cyclones chapter regarding Operation and Maintenance.

26.12 Potential Operating Problems Summary

Some possible operational problems are as follows:

1. Nonuniform gas distribution both into and out of the device
2. Excessive dust level in the discharge hopper
3. Insufficient dust level above discharge device
4. Air infiltration leaks at gaskets
5. Trickle valve or rotary lock leaks
6. Damage to cyclonic zone (body) of cyclone
7. Tube sheet leaks.

26.13 Optimization Techniques

Some optimization techniques to consider are as follows:

1. Provide smooth gas flow both into and out of the device
2. Carefully control the dust level in the hopper
3. Augment the sealing capability of the dust discharge device by maintaining a dust level above the discharge device
4. Maintain the integrity of the gaskets or seals on all access ports

5. Check and maintain the sealing capability of the discharge device

6. Make certain that the cyclone body is smooth and free of dents, weld scars, etc.

7. On multicyclone collectors, make certain that the tubes are properly seated onto a flat tube sheet.

26.14 Check Inlet and Outlet Ductwork for Uniform Gas Movement

Suggestions to explore:

Check the ductwork design both into and out of the device. If turning vanes were used initially, are they still present and functioning? If an inspection reveals disturbances or areas of solids build-up, perhaps a CFD model is warranted. With cyclones, the particulate (dust) tends to channel in a stream that may or may not be the same stream as the carrying gas. If the installation is new, perhaps using longer radius elbows will enhance the flow of gases to and from the device. Often some minor adjustments to the gas flow pattern can improve performance. If dust is building up in the gas inlet ductwork (primarily as the lower portions of the ductwork, cleanouts may be needed to get that dust out of the gas stream prior to the dry cyclone.

26.15 Inspect the Hopper If Used

In the zone above the dust in the hopper, the particulate environment might include fine particulate that tends to stay airborne combined with large particulate that wants to settle out. If the dust level can rise excessively, the distance between the captured dust and the discharge zone of the cyclone decreases. The fine particulate might tend to be pulled upward into the discharge of the cyclone(s). Thus, it is important to maintain an adequate separation zone between the dust and the cyclone(s). Various firms offer dust level monitors that sense the dust interface with the gas over the dust. The signal from such a sensor can sometimes be used to modulate the operation of a rotary lock or provide guidance for maintenance personnel to set the dump sequence of a dust discharge trickle valve. In some applications, the zone above the accumulated dust in the hopper is separately ventilated to a shaker or pulse type baghouse for filtration. The externally mounted filtration device discharges the filtered air back into the cleaned air plenum of the dry cyclone. The latter technique is often used where the designer anticipates that a large amount of fine particulate over the large particulate in the hopper can be expected (i.e., where highly friable particulate is suspected). The particles are so tiny that filtration is needed for separation and simple settling cannot be trusted to achieve the desired result.

26.16 Use Dust to Augment Sealing

As a corollary to the above, the dust sealing ability of the dust discharge device is often improved by maintaining a level of dust above the discharge device. The manufacturer can provide recommendations as to the depth. As above, a dust/gas interface level sensor can allow the proper control of the dust level.

26.17 Check Integrity of All Gaskets

All access doors and flanges should use functioning gaskets and seals. A dry cyclone operated under suction must be fully sealed to prevent ambient air infiltration that whose gas movement could lift dust out of the device and decrease the device's gas cleaning performance.

26.18 Double Check the Dust Discharge Device If Fitted

During maintenance inspections, the sealing integrity of the dust discharge device should be checked and repairs, if any are required, should be made. On systems operating under induced draft, a "smoke bomb" type test could be used to determine if air is leaking into the dry cyclone. If you want the best performance, there should be no leaks inward.

26.19 No Dents!

Any damage to the body of a dry cyclone should be repaired. During operation, the cyclone body locates the path that the gas stream follows as it spins. Any bump, weld scar, seam etc. can upset this pattern and cause a reduction in particulate removal performance.

With multicyclone type collectors wherein numerous cyclone "modules" are mounted on a tube sheet, any gaps between the cyclone modules and the tube sheet should be eliminated otherwise dust can pass through without being separated. If modules are removed, the tube sheet should be inspected for flatness and any repairs be made prior to replacing the cyclone module.

27

Absorber Optimization

27.1 Define Complete Absorber Operation Including Ancillary Equipment

In a packed tower, the packed bed design and operation provides the basis upon which the desired mass transfer of the contaminant gas into the liquid can be achieved. The packed section (or "absorption" or "contact" section) is essentially of mechanical design (the surface of the absorbing or cooling liquid being mechanically increased, or "extended" as that liquid passes over the media) but that often exhibits a "chemical" function. As the gases are absorbed, for example, there may be an exotherm that could reduce the absorption of subsequent gases through the reduction in gas solubility. The absorbed gases may react with chemicals in the liquid thus producing reaction products that may reduce the rate of absorption. Those reaction products may also produce solids and/or scale that can mechanically affect the flow of the liquid over the packed bed media. As a result, various types of packing may be harnessed by the designer to maximize the mass transfer while minimizing solids build up and scaling. The options range from dumped type to structural.

A more subtle but no less important condition can occur wherein the pH and temperature of the liquid in the contact bed is not optimized.

For example, with the absorption of gaseous ammonia into an acidic solution, many packed towers are applied using the "conventional" arrangement of using a counterflow configuration with the recycle sump at the packed tower base. Sometimes, the recycle sump is even external to the absorption section. Make up acid under pH control and make up water under level control are often added to this sump. From the sump a pump recycles the liquid back to the top of the packed tower and the liquid is distributed within the tower using sprays or distribution troughs. The liquid passes through the absorption zone and drains back to the sump. Some of the liquid is typically bled away to remove the reaction products (in this example, salts). This mode of operation is often adequate however the procedure is too often not optimized.

FIGURE 27.1
Structured Packing (Brentwood Industries, Inc.)

27.2 Minimize Potential Problems

Some problems are the following using ammonia absorption as an example:

27.2.1 Unless the Gas Is Absorbed and Reacted, Minimize Recycle

The recycling of the liquid returns some of the products of reaction back to the absorption zone could inhibit the subsequent absorption of the ammonia. The intent is to keep individual stages focused on their specific function rather than needlessly recycling liquid back to the gas/liquid contact area.

27.2.2 Don't Waste the Make Up Water

Since make-up water is usually added to the sump then a short time later is bled away, water can be wasted. Arrange the piping so the makeup water goes through the absorber at least once. This often involves adding the makeup water directly to the recycle header.

27.2.3 Control Temperatures

In an exothermic reaction, much of the cooling effect of the make-up water is lost since a portion of that water is bled away before the water can control the exotherm in the absorption zone. Conversely, if the reaction is endothermic, temperature control using heated infeed liquid may improve the absorption process.

27.2.4 Careful pH Control, If Used, Is Critical

The pH in the absorption zone is not optimized. Ammonia scrubbers using acid are often operated at a pH of 3 or less. Higher pH levels may result in the gas phase reaction of the ammonia with the acid causing the gas phase formation of particulate. That particulate may cause a visible emission. If the pH is measured in the scrubber sump, the pH is post reaction. The liquid surrounding the pH sensor contains both the unreacted acid, products of the reaction (salts), and other compounds that may affect the pH reading.

27.2.5 Make Chemical, If Used, Pass at Least Once Through

If the make-up acid, for example, is added to the sump and that sump is bled some of the unreacted acid will be bled needlessly (and perhaps expensively) away prior than passing through the absorption zone where the acid is useful.

27.2.6 Locate Sensors for Most Rapid Response

Given the typical high liquid volume of the sump (often sized to maintain the proper pump suction pressures), a pH change response time delay is inherent. A large volume of liquid surrounds the probe. The absorbed ammonia tends to raise the pH of the sump, but a large sump requires large amounts of ammonia to significantly alter the pH. If in addition the pH is controlled near pH 7 (often required to meet liquid discharge permit limits), the response parameters of the probe may be at or near the non-linear near vertical pH response curve range under which the system response creates reduced control. If make up acid is added to this large tank volume, the acid must be thoroughly mixed to obtain a representative and stable pH. These conditions conspire, though subtly, reduce the overall system response time if the ammonia infeed rate changes.

27.3 Adjustments to Consider

To optimize the absorption, certain adjustments can be made:

27.3.1 Use Dilute Chemical Make Up If Possible

To reduce the negative influence of the products of reaction, the liquid entering the liquid header(s) at the discharge of the absorption zone must be kept as dilute as practical. Adding the make-up water just prior to the liquid distribution header can achieve this.

To conserve water, the liquid circuit can be configured to only blow down reaction products. This can be achieved by configuring the recycle loop by making the water at least go through the absorption zone once. Adding the make-up water (as in "a" above) also conserves water.

27.3.2 If Exotherms Is Present, Add Cold Make Up Water to Header

As a bonus, adding the (typically colder) fresh water make up at the header rather than at the sump can also help remove some of the reaction exotherm. Adding the colder liquid at the infeed header rather than at the sump provides the desired heat transfer at a location (at the discharge of the cleaned gas(es) where that cooling can be most effective. Doing so can thereby stabilize the absorption.

27.3.3 Premix Additive Chemicals

To optimize the pH in the absorption zone, the make-up acid can be added and mixed using an in-line static mixer prior to injection into the scrubber liquid distribution stage and therefore into the absorption zone. Adding any make up chemical to the liquid header rather that into the sump makes that chemical go through the absorption stage at least once. If the products of reaction are then bled from the sump, the flow of those reaction products back to the absorption zone where those products may inhibit absorption is reduced.

Historically, premixing is accomplished by injecting the chemical into the pump suction. The pump acts as a mixer. Some of the downside effects are that the chemical is typically strong at that point and localized corrosion and/or erosion may occur. If the chemical reaction creates a gaseous reaction product, cavitation may also occur though it is difficult to detect. Plus, the uniform distribution of chemical is properly needed at the liquid distribution section of the packed tower (spray headers, troughs, etc.) and may not be uniform at the pump.

Alternate pre-mixing involves purposely creating changes of direction in the piping prior to the liquid distribution stage of the tower. Multiple 90-degree elbows, though adding to the head losses, are often effective. Inline static mixers such as shown in the following figures can also be effective and take up less space.

In these devices, fixed internal baffles cause the chemical and carrying liquid to mix though the liquid stream travels in essentially one direction.

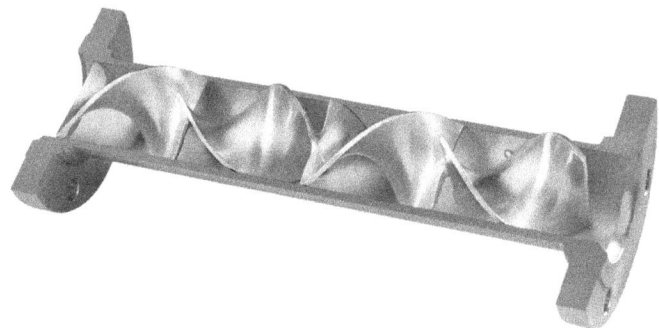

FIGURE 27.2
Kenics Static Mixer photo.

A typical configuration is shown in Figure 27.2. The internal baffles that cause the mixing are shown in Figure 27.3. These devices can be used singly or in series depending upon the mixing difficulty.

Another mixing method to provide uniformity to the scrubbing liquid infeed stream is to use a mix tank equipped with an agitator. Sometimes these tanks are equipped with a dedicated pump that draws a liquid stream from the tank sump and returns it to the top of the tank on a continuous basis. A portion of that recycle is bled on demand to the scrubber liquid recycle circuit.

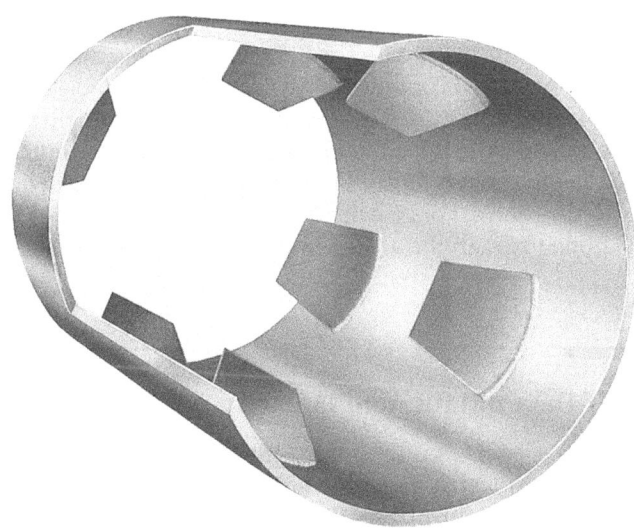

FIGURE 27.3
Kenics Photo showing internal baffles.

27.3.4 Choose Effective Probe/Sensor Locations

A pH probe mounted in the liquid inlet header (or in a service by-pass which-ever is most convenient) can produce superior pH control at the top of the absorption zone. The signal from the pH probe is used to control the make-up acid amount.

The relatively low volume of liquid in the inlet header provides for a faster pH response than if the pH probe is mounted in the higher volume sump.

To reduce operational complexity, the existing device level control and blow-down hardware (often using a flow meter and control valve on the pump discharge) can be retained. The sump pH probe can be used to monitor and control the blowdown since a representative blowdown pH is often needed (or required by the operating permit) if that flow is to be sent to sewer or other discharge.

27.3.5 Seek Uniform Liquid Distribution

As mentioned in previous chapters, the liquid needs to be properly distributed across the absorption zone. Sometimes adjustments to the liquid header(s) or liquid distribution device are needed.

The added hardware for optimization is usually minimal. How the hardware is used is the key to effective optimization. The make-up acid and make up water are pre-mixed, and pH controlled. The make-up stream is then blended with the recycle from the sump. The liquid distribution stage is the same. The liquid still drains from the absorption zone into the sump or sump tank. The sump pH probe can still be used but its chore is to monitor the post reaction pH (same as the blow down pH).

An added effect is that the system will now more rapidly respond to any changes in the ammonia input rate. The infeed pH is measured in a much lower volume, flowing, premixed condition away from the less stable zone near pH 7.

If the above optimization is implemented, the gases moving through the absorption zone will be exposed to liquid that is the most dilute, of the optimum pH and is the coolest. Every volume of make-up water will have gone through the device at least once prior to discharge. Every weight of acid will have at least gone through the absorption zone once.

27.4 What If the Containment Is Easily Stripped?

In some processes, such as odor control or the absorption of alcohols, the contaminant can easily strip from the absorption liquid. If the contaminant vapor pressure is not neutralized through chemical reaction and/or cooling, recycling that liquid to the absorption zone can cause the contaminant to strip from solution causing a reduction in the performance of the hardware.

For odor control, often pH control and oxidation are typically used to reduce that vapor pressure by converting the absorbed odor to an oxide or salt. The odor may be relatively soluble compounds such as hydrogen sulfide or perhaps lower solubility such as reduced sulfur species. The odor may be amines or other soluble odorous organics. For the absorption of alcohols, high bleed rates (water in then directly out) are sometimes used. For the latter, additional cooling and/or multiple stages are incorporated so that the partial vapor pressure of the liquid at the point of discharge (the device outlet) is kept at a minimum.

27.4.1 Possible Problems Given the Chemical Addition Odor Control Example

With odor control, for example, the emissions control can reveal various problems:

1. The addition of chemical may cause an odor. Some oxidant chemical such as sodium hypochlorite is often received "stabilized" at an elevated pH. For use, the pH of the stabilized chemical is reduced prior to injection into the odor control device. Reducing the pH activates the oxidation potential however reducing the pH too fast or at the wrong location can free undesirable "bleach odor" and waste chemical.

2. If an oxidant such as hydrogen peroxide is used, the peroxide may react slower than desired thus reducing the odor reduction of the hardware.

3. If the chemical is not mixed efficiently, erratic odor reduction can result.

4. The residual oxidation/reduction potential (ORP) of the post reaction residual (usually measured in the sump) may be insufficient for the application. Often an excess ORP is required to move the reaction in the desired direction rather than simple equilibrium.

27.4.2 Problems If the Absorbed Containment Easily Strips

With the control of alcohols and similar compounds that easily strip, some problems can be as follows:

1. Excessive water consumption.

2. High liquid discharge temperature that raises the partial pressure of the contaminant causing an unwanted release of the contaminant as vapor.

3. Buildup of slime or scale in the absorption zone.

4. Higher than optimal (low) liquid concentration at the point of cleaned gas discharge of the control device.

27.5 Optimizing Odor Control Applications

To optimize an odor control system, some of the following techniques may be considered:

1. Premix and pH control the additive chemical (if that chemical can impart its own odor) prior to injection of the chemical into the recycle liquid stream. Much like the general recommendation (above) for an acid-duty packed tower, premixing and establishing better control of the chemical as that chemical enters the absorption zone can improve performance and reduce chemical costs. If adding the chemical to the sump can be avoided, consideration should be given to premixing to optimize the performance.

2. To achieve the required odor reduction using a chemical such as hydrogen peroxide, adequate retention time of the oxidant in solution may be required. Peroxide suppliers publish data that indicates typical retention times to complete the oxidation. Some soluble odors can take many minutes for the oxidation to go to completion. As a result, often the device sump (whether internal to the device or remote) is inadequate for the current conditions. The device retention sump or tank may at one time have been adequate but given process changes no longer is. Some systems started with only a sump large enough to meet the hydraulic demands of the recycle pump (volume and NPSH). In that case, an additional sump may be required to achieve the required residence time. If a tank exists but is too small and additional tank (space permitting) may need to be "sistered" with the existing tank. For systems that already use an external retention sump, adding the peroxide to the device drain can sometimes be adequate. Another technique to increase the residence time in an existing tank is to add a center baffle ("mid-feather") that forces the oxidant to take a longer path than if the oxidant was simply added to the sump. Assuming proper liquid coverage in the absorption zone, sometimes decreasing the liquid recycle rate to the device serves to increase the in-sump residence time for oxidation.

3. Improved mixing of the sump or external recycle tank can improve performance. Paddle mixers are possible candidates to provide the mixing. Another method involves "turning the tank over" by drawing liquid from the bottom of the sump or tank and returning some of that flow back to the tank. If the main recycle tank is of adequate capacity, a portion of that flow can be sent back. An additional small pump might be needed. Typically, these additional pumps only "turn over" about 10% of the sump volume.

4. Often an excessive amount of oxidant must remain in the sump to maintain the oxidation reaction. Though some of this oxidant will be wasted by way of the blowdown, the excess is required to drive the oxidation reactions. Thus, an ORP probe mounted in the sump or blowdown header is often used to provide the control signal. Based upon experience learned for that specific application or device, the ORP probe and controller provides the signal that administers the oxidant chemical. On hypochlorite type systems, the reaction is rapid and a more neutral ORP signal may be adequate. If that signal is not reliable, a specific ion electrode could possibly be a substitute. For peroxide-based systems, the ORP may be set at a higher potential. The goal, of course, to conserve chemical is to only set the ORP set point at a level required to achieve the desired odor reduction.

27.6 Optimizing Applications That Easily Strip

For applications that are designed to control contaminants that easily strip from solution, there are some techniques that may be applied:

1. Excessive water consumption is common on systems that are configured for "water once through". The facility may have available to it copious amounts of water or "used to have" copious amounts of water. The simplest operating method was to send the water into the scrubber inlet header then directly out. With strippable type compounds, the liquid rate (concentration of contaminant in the liquid), the liquid temperature, and the location of application in the control device are the functioning items that define the optimization possibilities. If excessive water is being observed (or measured) then reduction in temperature of that liquid to reduce the vapor pressure, reducing the contaminant concentration and the location of application of the cleanest liquid is paramount.

2. Sometimes reducing the infeed liquid temperature can help achieve optimization. Even if the operation is "water once through" that infeed water may be excessive. Removing some of the heat using heat exchangers on the gas stream, liquid stream or both can improve performance. Many areas are limited by the ambient wet bulb temperature (if cooling towers are used) for defining the temperature to which the liquid stream can be reduced. Chillers can be added if the cost justifies their use. If the gas stream source contains water vapor, sometimes the condensation of that water vapor to liquid water can

serve to dilute the discharge stream thus reducing, at least to a small extent, the partial vapor pressure of the contaminant.

3. Some of these strippable contaminant applications are "organic" related. They may include fermentation processes or baking processes that could evolve alcohols for example. These type sources can lead to slime or scaling buildup that can coat transfer media (such as packing) in the absorption zone of the device. The coating can reduce mass transfer thus reducing performance. A switch to larger free-pass type packed bed media run at a higher liquid to gas ratio may improve performance at modest additional cost. Packed tower media vendors may suggest larger dumped type packing or even structured type packing as an alternate.

4. The interface conditions at the junction of the liquid and contaminant gas at the point of exit from the control device determines the ultimate gas outlet conditions. If strippable contaminants still exist at the gas outlet, some of those contaminants may still strip off. To reduce the tendency to strip, the most dilute, coolest, cleanest liquid must be applied at the point of gas discharge. Some devices applied to this type process use mesh pads for droplet control. The mesh pad inherently builds up liquid until that liquid accumulates enough to be heavy enough to drain. That means that some contaminants also build up in the mesh pad. As they concentrate, they can reach equilibrium with the gas stream thus setting a "floor" below which the contaminant concentration cannot be reduced. It is sometimes possible to change the mesh pad to one that retains less liquid (larger, more open mesh) so that less liquid is retained. In addition, that pad can be sometimes continually face sprayed with cooled, clean water. One must make certain however that the mesh pad does not flood so consultation with the mesh pad supplier is recommended. An alternate is to add a separate cross flow droplet separator using clean, cooled water applied to an inclined chevron type droplet eliminator. The droplet eliminator holds up a minimal amount of liquid therefore the resulting liquid film is of lowest contaminant concentration thus the vapor pressure is lowest. The inclination assists in the draining of the chevron blades thus minimizing the liquid hold up and keeping those surfaces cool and dilute. These type units have been successfully used on ethanol-type processes.

28

Pre-Formed Spray Scrubbers Optimization

28.1 Balancing Absorption and Particulate Removal

It is easy to dismiss the pre-formed spray type scrubber as simply a cyclonic separator outfitted with spray nozzles. It is more complicated than that.

This type scrubber has two-design situations that need to coexist. Small droplets are needed to achieve the desired particulate removal and gas absorption while larger droplets are needed to provide adequate liquid separation from the gas stream. As a result, these type scrubbers incorporate two zones distinctive zones of operation. One is the mass transfer zone, and the other is the droplet separation zone. To complicate things further, these zones sometimes overlap.

Pre-formed spray scrubbers vary in size however the primary functional zone is where the nozzles and nozzle assemblies are located. The largest diameter portion of the design is for separation of droplets and for a lesser extent, gas absorption. Figure 28.1 shows a pre-formed scrubber installation.

To optimize mass transfer (as with other wet scrubber designs), the general goal is to provide an adequate surface area of liquid to afford the mass transfer. For economy, that surface area is usually applied at the minimal vessel volume (usually the gas inlet area). To optimize separation, however, the gas and liquid droplets are separated in a larger vessel wherein the distance between the droplets and gas is increased (the droplets are thrown to the vessel wall). Sometimes these requirements are in disharmony. The designer usually separates the problem by focusing on the mass transfer (upfront) zone since performance guarantees usually are met or missed given the operation of that zone. The droplet separation is often relegated to simple centrifugal separation however the performance of the mass transfer zone can have a direct effect on the separation zone. To optimize, these zones must perform in harmony.

Some things can be done however to optimize a pre-formed spray scrubber that is not functioning at the desired level of performance.

FIGURE 28.1
Preformed spray scrubber. (Bionomic Ind.)

28.2 Some Problems to Consider

Problems may include the following:

1. Poor gas absorption
2. Poor particulate removal
3. Insufficient droplet separation
4. Excessive power input.

28.3 Optimization Techniques

To optimize the performance, some techniques can be applied:

1. As mentioned above, the mass transfer zone determines the gas absorption performance of this type scrubber. The reasons are many. In this zone (at the inlet of the scrubber) the ratio of the surface area of the liquid to that of the gas is greatest. In addition, the "density" of the liquid droplets is also the greatest. This means that the probability of the contaminant gas into the liquid is also the greatest. After the gas stream leaves this zone, the high surface area droplet regime is forced (thrown by centrifugal force to the vessel wall) converted to a low surface area film of liquid. That film of liquid drains to the scrubber sump however its role as a gas absorption media is effectively ended. Thus, to optimize gas absorption, one must endeavor to provide the greatest surface area per unit volume at the mass transfer zone. Obviously, using a finer liquid spray in that zone will achieve that result. The caveat, however, is that if the spray generates droplets that are too small for centrifugal separation, liquid entrainment from the device could occur. The nozzle selection thus becomes critical. To optimize, sometimes the nozzles can be changed to ones that provide a denser spray of liquid but whose residual droplet size is greater than about 200 microns so that those droplets may be separated efficiently. Nozzle suppliers can be consulted regarding possible nozzle substitutions. Also, quite often the nozzles used are "fan spray" types that provide a flat spray pattern. This pattern typically relies on the shearing action of the moving gas to break up the liquid into smaller droplets. An additional header can possibly be added to supplement the fan sprays that use a full cone spray pattern that provides a greater density of smaller droplets. Another improvement could be increasing the residence time of the spray

in the mass transfer zone. More contact time can improve the mass transfer particularly if the absorbed gases react with chemicals in the liquid spray. If the scrubber, say, uses a 90-degree gas inlet, it may be possible (though likely not inexpensive) to change the gas inlet to an involute type that extends perhaps 180 degrees around the vessel. A further improvement for gas absorption may be to add wall boxes outfitted with supplemental sprays aimed in the direction of the cyclonic rotation of the gases in the separator. These sprays must have a residual droplet size large enough to be affected by the cyclonic action of the separator (i.e., large residual droplets are needed if wall box sprays are used. In general, one can improve the mass transfer to an extent limited by the residual droplet size of the applied spray.

2. For particulate removal, the "standard" pre-formed spray scrubber is configured to have the gas and liquid moving in the same direction. This arrangement is simple and adequate particulate removal can be achieved. As learned in other sections of this book, one wants to impact the particulate into the liquid. Thus, what is really needed is that the target particle be moving faster than the liquid droplet. Though large (above 10-20 microns) particles can be removed using the concurrent method, to boost the removal of finer particulate may require the application of a counter-current spray pattern. Sometimes a full cone spray pattern can be applied "into the wind" to increase the relative velocity between the particle and droplet. This header (or a plurality of headers) might be able to be fitted into the mass transfer zone. If not, perhaps an additional "pre-scrubber" zone can be added ahead of the gas inlet. This technique has been applied successfully to dissolving tank vent scrubbers in the pulp and paper industry. As mentioned above, the caveat is that the residual spray droplet size must be large enough to be cyclonically separated (too fine a spray may not separate). Yet another optimization technique to consider for particulate removal is to, if possible, increase the pressure drop on the scrubber. The best place to increase the pressure drop is often the area just before the gas stream enters the cyclonic separator. The reason is that the droplet density is greatest at that point thus the probability of a particle being impacted into or being intercepted by a droplet is greatest at that point. Assuming the prime mover can overcome an increase in pressure drop, adding a baffle (fixed or movable) at that point can provide the required increase in pressure drop. The original scrubber vendor or perhaps a consulting engineer can provide the approximate open area required to achieve the pressure drop increase.

Insufficient droplet separation often occurs if the interior surfaces of the vessel are too rough and allow for liquid accumulation and entrainment or the incoming droplets are too small to be centrifugally separated. As mentioned above, if the applied spray droplet size in the mass transfer zone is too small then those droplets may avoid separation. The remedy if that is the case is to "tune" the mass transfer zone spray back to generate slightly larger droplets (usually the spray droplet size decreases with nozzle pressure). If decreasing the flow does not result in better liquid separation, then a change in nozzle(s) may be required. An additional technique to investigate involves adding a chevron type separator module to the vessel. One must have adequate vessel height for its installation, however, and the face velocity into the chevron must be within the chevron module vendor's requirements (often about 8–10 ft/s) and any swirling of the gases must be stopped. Sometimes the upper part of the cyclonic separator can be increased in height to accommodate the chevron. If a high solids environment is expected, often clean in place headers that spray the module are included as part of the optimization. A further improvement could be the addition of a crossflow separator after the cyclonic separator. These crossflow devices use chevrons (either vertically oriented or inclined on an angle) to separate the entrained droplets from the gas stream. Suppliers such as Munters and Coastal Technologies provide such devices.

3. Excessive power input usually is a result of the scrubber pressure drop. In a pre-formed spray scrubber, the pressure drop primarily occurs as the tangential location where the gas stream enters the cyclonic separator. This pressure drop, however, is required for proper particulate removal and that removal may be part of a permit to operate and/or a performance guarantee. To optimize the pressure, drop, the original scrubber vendor should be contacted to see what pressure drop is the lowest that can be used to meet any contract or compliance requirements. Once that pressure drop is known, the open area at the entrance to the separator can be calculated and that area be adjusted to suit. Also, if excessive liquid is being applied to the device, the pressure drop may, in turn, be excessive. As above, the device vendor should be consulted to determine the lowest acceptable liquid rate. Adjustments to the liquid rate on this type device may provide a pressure drop adjustment of about 5%.

29

Precipitator Optimization

29.1 The Requirement of "Synergy"

To achieve the highest performance at lowest cost, few air pollution control devices require more "synergy", more "balance" between the primary and ancillary components than precipitators. The "primary" components are the vessel, the "can", and the ancillary components are the devices used to make the "can" work.

Whether a dry or wet design, these components must all work together. Changing a process or moving a precipitator to a new or modified process often requires going back to the beginning. For a dry application, resistivity testing of the particulate may be required. If acidic or other corrosive gases are now part of the input contaminant mix adjustments may be required. For wet electrostatic designs, the inlet particulate loading, humidity, or settling characteristics have changed, the "problem" may have to be redefined.

For these reasons and others, to optimize the performance of a precipitator, it is suggested, by this author at least, that you seek the advice of the original device supplier or enlist the engineering expertise of consultants in that field. Investing in a thorough study of the problem and focused definition of the project goals up front, increases the likelihood of project success. There are a variety of sources, at the time of this writing, of locating experienced precipitator consultants. One combined source is the www.apcnetwork.com.

29.2 Primary Precipitator Components

Why the need for a study? Take the "can" velocity. In a dry precipitator, during the rapping of the collecting plates, the particulate is discharged and hopefully falls to the collection hopper. Given the impact energy of the rappers, plate spacing, particulate aerodynamic characteristics and can gas velocity, some of this particulate may be entrained and leave the collector. Precipitator designers carefully design the vessel based upon these characteristics and choose the vessel dimensions so that the local gas velocity is below the carrying velocity of the particulate. Change the particulate characteristics and/or the treated gas volume, and adjustments to the vessel may be needed. For a modular precipitator, perhaps a new module may be needed. For major changes, an entire new device may be the best solution.

29.2.1 Gas Inlet Ductwork

Say the inlet gas ductwork had to be changed and the device performance deteriorated. These devices, wet or dry, for best operation require uniform inlet gas distribution. Inherently low in flange-to-flange pressure drop, energy wasting changes of gas direction are minimized. But minimization does not mean elimination. Often perforated plates, turning vanes, or other such devices are needed to achieve acceptable inlet gas distribution. The goal is to ensure that the internal collecting surfaces, wet or dry, are functioning as designed.

29.2.2 Importance of Particulate Characteristics

Changes in the particulate characteristics may also force a re-definition of the problem. The particulate's characteristics may need to be modified through inlet gas conditioning. Though typically less important in the use of wet precipitators, inlet gas conditioning may be required to enhance the collection characteristics of the particulate. So called "dry fog" inlet gas conditioning may be required wherein atomized water is administered under controlled conditions wherein the fog dries to completion (minimal residual droplets). Sometimes specific chemicals are added to the spray based upon particulate characteristics testing.

After the particulate is captured, it must be removed from the device. It must be removed before it causes trouble. In a dry precipitator, the control of the dust level in the hopper is particularly important. Excessive particulate may entrain out of the device. If the particulate is hygroscopic, local buildup and plugging could occur. These hoppers are equipped with dust-level sensors such as a Bindicator shown in Figure 29.1.

FIGURE 29.1
Bindicator.

The setpoint of the highest level of particulate may need to be tuned to minimize the entrainment of particulate. Often an internal inspection of the device will reveal traces of dust carry-over. The lowest particulate level may impact the performance of the particulate removal devise (often a rotary lock). To be properly "fed" some locks enjoy the use of a small inventory of particulate above the lock. Some applications, prone to build up or "bridging" require the zone above the discharge device to be kept free. Local disrupting devices may be needed in those applications.

29.2.3 Collecting Surface Characteristics

For wet precipitators, if the water is in whole or in part recycled as a portion of the collecting surface irrigation, the setting characteristics of the collected particulate could be significant. Often the lowest irrigation water speed is in the device itself and not in the irrigation recycle loop. Some wet electrostatic precipitators use internal distribution weirs. If use of a weir causes the water velocity to fall below the setting velocity of the irrigation water, local buildup can occur. Further separation of recycle solids through settling external to the device and/or increases in makeup water to the internal distribution network may be required.

29.2.4 Electrode Type and Characteristics

The internal electrode design, location, and function are of critical importance. Individual precipitator supplying firms have developed, modified, and applied electrode design configurations that they feel best suit the performance needs. Some after-market suppliers have also tried to make improvements in electrodes. Caution must be taken, however, if the precipitator is operated under a Permit that explicitly defines the device. In that case, it is suggested that you contact the original supplier. If the Permit simply requires periodic testing or continuous monitoring, you may be able to change the electrodes or other internal components if the performance is maintained or is improved. It is best to check with the local regulator agency if these changes are contemplated.

29.3 Primary Components

Some of the Primary Component requirements are quite simple but are overlooked or change with time.

Is the device installed so that the internal components are vertical? Sounds simple but these devices can settle over time. Dry or wet, the internal devices "hang". In a precipitator the electrodes (even if restrained) tend to hang. In a dry device, the electrodes may be weighted but they need to be in the proper orientation to the collection plates to maximize performance. In a wet device, the tubes need to be truly vertical so that, in similar fashion, the distance between the electrode and the wetted tube is maintained within the design distance parameters. A double check, and adjustment if needed, of the verticality of the device may be required.

29.3.1 Ancillary Components

The ancillary components for this discussion include the rapper design and application for dry precipitators; the irrigation system for wet precipitators, and the power supply for both.

As mentioned earlier, synergy is needed for success. For dry precipitators, the original supplier and/or a qualified consultant is suggested to be retained for any investigation of the rapper design, operation, or adjustment. Though the rapper section is in the vessel, it may be helpful to view them as ancillary. Some adjustments can be made to improve performance if done under the watchful eye of the original supplier or experienced consultant.

For a wet device, the irrigation system can often be refined. If the irrigation is by city or plant water, it is likely that few adjustments are needed. Sometimes if flow monitoring is lacking, flow instrumentation and control is added. Doing so helps minimize water consumption and helps identify and solve local buildup issues in the vessel. If the irrigation water is recycled in whole or in part, solids characteristics and concentration may need to be addressed.

Often, increasing blowdown can reduce solids concentrations to acceptable levels. Settling characteristic tests can be conducted to estimate the settling rate if the recycle is held in an external tank prior to injection. The tank may simply be of improper size and may need agitation. If fresh make up water is added to the tank, it may be best, instead, to add the water directly into the recycle header (thus diluting the recycle and forcing the "clean" water to at least take a once-though path through the precipitator). In some cases where the solids have value, it may be best to invest in a higher efficiency solids reclamation device. The existing system may just use settling or a liquid cyclone so filtration may be a substitute to consider and offer both recovery and performance advantages.

29.4 The Power Supply

For both wet and dry, advancements have been made and are being made in the power supplies that provide the high voltage to the electrodes and that control that voltage. Improvements in the control electronics allow the

FIGURE 29.2
NWL power supply.

more modern power supplies to provide the high voltage more uniformly this improving collection efficiency.

High frequency power supplies as provided by NWL and others might be considered as a substitute and improvement over older or lower efficiency devices. The NWL design unit is shown in Figure 29.2.

These newer power supply devices apply improved voltage regulation and stability thus maintain the high voltage application within a narrower range. The devices require considerable investment thus their need requires detailed investigation and study.

After design parameters of the application have been defined and any updated testing has been completed, changes to the Primary Components (and related costs) can be obtained followed by an analysis of the need for any upgrades to the Ancillary Components. Properly applied, the precipitator should show improvements in performance through optimization of these critical components.

30

Quencher Optimization

Probably, the most severe environment that air pollution control equipment must resist occurs in high-gas temperature applications such as hazardous waste incinerators. Gas inlet temperatures can exceed 2200 degrees F (1200 degrees C). The combination of corrosive gases, high gas to liquid temperature differentials, rapid water evaporation rates, and the rapid near explosive flashing of water to steam all combine to create an exceedingly aggressive environment. It is not uncommon to hear from nearby the hostile environment going on in a hot gas quencher.

The quencher must operate reliably. Maybe the word should be must. Not only does the quencher reduce hot gas temperatures in preparation for gas cleaning but also likely protects expensive downstream equipment from thermal damage and/or destruction. Most quencher designs are therefore configured to be reliable. It is not uncommon that the quencher is over-designed mechanically (incorporating conservative gas velocities, residence time, and using excessive water for example) and its associated equipment such as piping, and water source is sized and selected for reliability. It seems simple. Just cool the gases with water. But it is not simple. In a quencher, the cooling is not immediate. There is a time delay factor (much like hysteresis in magnetism) as the water, whether injected as a spray or in sheets, absorbs heat, increases in temperature, and changes phase (from liquid to vapor). This all can occur within a fraction of a second but not instantaneously in the quencher.

30.1 Possible Problems

Conservative design aside, some problems can occur with quenchers that often are revealed only after the installation. Some of those possible problems are as follows: (the following assumes a vertically mounted quencher)

1. Insufficient gas cooling (higher than desired gas outlet temperature)
2. Spray drying to particulate
3. Poor gas distribution causing localized "hot spots" or refractory wear

FIGURE 30.1
Quencher, Bionomic Industries.

4. Inability to handle peak temperature spikes

5. Thermal expansion issues causing structural problems

6. Local corrosion issues

30.2 Optimization Techniques

Some optimization techniques to investigate include:

1. If insufficient cooling is observed and the quencher's size (usually length) cannot be changed given lack of space for a new one, the focus turns to the liquid circuit. With quenchers, the liquid injection

amount and method are critical. The quenching rate, like absorption, is surface area dependent. The liquid's surface area is extended so that the water can convert to water vapor rapidly. The designer typically selects nozzles and spray patterns to provide the required liquid coverage in the quencher. A uniform gas flow pattern is usually assumed. The liquid flow rate (volume) is selected to saturate the gas stream at least adiabatically plus a safety factor. Some of the water will be sensibly heated but not participate significantly in the conversion to water vapor. That safety factor flow leaves as heated water. It is the droplets that heat and then evaporate to near completion that do the work. The nozzle droplet size is often determined to prevent the droplet from drying completely (more on that subject follows below). Reality can step in however requiring a change in the nozzle type, spray pattern and droplet distribution. Sometimes turning a spray header upward and changing to a full cone nozzle can both increase gas/liquid residence time and provide additional cooling without increasing the quencher vessel size. In addition, the quenching can often be "staged" by using multiple headers or altering the flow rate to existing headers through external piping changes. "Staging" injects the higher volume of water at the upper portion of the quencher wherein the evaporation rate is greatest. The goal is to drop the gas temperature rapidly to below about 500 degrees F (260 degree C) wherein the evaporation rate per unit volume tends to decrease. The decrease occurs because the gas surrounding a droplet already contains significant water vapor. "Staging" takes advantage of the high evaporation rate of that hottest zone. "Staging" also rapidly reduces the gas velocity so the residence time in the quencher is optimized. Nozzle suppliers such as BETE Fog Nozzle, Spraying Systems and others can help select alternate nozzles and perhaps recommend changes in the spray location whether staging is used. Experimenting with a different set of spray nozzles is often worth the effort prior to investigating more expensive and difficult options. The nozzle vendors do not, however, guarantee results since there are too many operating variables that ultimately determine success. You would need to provide operating data such as the existing nozzle size and type, location, and process conditions (gas inlet and outlet temperature, pressure, and humidity) to their applications professionals. A drawing showing the quencher header size and location would also be needed. If the ductwork allows it, sometimes a high pressure hydraulically atomized or an air-atomized spray can be added ahead of the quencher to provide additional cooling prior to the quencher. But with a caveat…

2. If the droplets are too small and contain dissolved and/or suspended solids, spray drying of those solids into gas borne particulate can occur. This drying occurs where the differential in gas to liquid

temperature is the greatest. At that point, the heat transfer is greatest as is the evaporation rate. Thus, if a spray is added or nozzles are changed, it is best to apply the cleanest, lowest dissolved solids water at the hottest zone of the quencher. Quite often, the optimization involves using softened water or stripped condensate to the highest temperature liquid injection point. Some facilities blend recycled water with softened make-up water at that point to reduce spray drying. Sometimes adding a filter for that portion of the quencher circuit can provide enough suspended particle reduction. It is helpful to perform a water analysis to determine if filtration alone can help. If the dissolved solids are minor and the droplet size is large enough, the spray drying will be minimized. Thus, filtering the water and shifting to larger droplets may be what is needed for optimization if the dissolved solids are low.

3. Poor gas distribution at the inlet of a quencher is all too common. Often the gas inlet ductwork includes an elbow or other direction-changing elements are located ahead of the quencher. To reduce costs, these typically refractory lined pieces are made to a minimum size and are located as close to the quencher inlet as possible. In addition, there may be expansion joints or the like that can upset the gas flow pattern. If the hot gases are not distributed evenly in the quencher, the quencher liquid circuit (usually comprised of spray nozzles) is not optimized and hot spots can be created. A thermal imaging scan of the quencher can help locate those hot spots. Local passageways of hot gases can occur thus resulting in higher than desired gas outlet temperatures. Usually at least two duct diameters of straight ductwork are required ahead of the quencher gas inlet to provide acceptable gas flow patterns. Turning vanes are usually not an option given the cost of applying such vanes in such a high temperature environment. If the inlet ductwork cannot be raised to create a uniform gas inlet pattern, the options are limited and usually expensive. A CFD model can be run on the quencher spray pattern and nozzles can be relocated to place the greatest spray in the hottest gas stream. Some optimization methods have included adding a "core buster" disc in the center of the quencher at a point below the uppermost spray injection point. The "core buster" makes the gas stream divert around the disc thus acting as a gas distribution device. The residual spray from the uppermost liquid injection area falls onto the disc thus enhancing the cooling. The disc, however, must be supported from the quencher wall and must be resistant to both the corrosive environment and the temperature extremes. The disc need only reduce the quencher area by less than 10% or so. The disc just provides some resistance to gas flow that aids in stabilizing the gas flow. The pressure drop on the quencher however

will increase. The quencher supplier, if they recommend a disc, will need to calculate the pressure drop increase. The capacity of the prime mover would then need to be checked to see if the addition of a disc will allow adequate process ventilation. As can be gleaned from the above, it is best to provide the 2+ duct diameters above the quencher inlet.

4. Some processes cruise along in a controllable steady temperature condition while others exhibit temperature spikes. These spikes are often unpredictable and could cause damage to the quencher and related equipment. Often, if spikes are expected, the liquid circuit of the quencher is oversized (excessive water flow continually) or an additional emergency header is used. To keep the emergency header cool, a constant flow of water at a minimum is provided. If the spike occurs "on schedule" the emergency header flow is switched to full flow prior to the "event". Often two solenoid valves are used in parallel with a small one providing the constant flow and the larger one opening when a spike is expected. If the spikes are random or cannot be predicted and an oversized liquid flow is not practical, temperature sensors located at the point closest to where the temperature spike could occur a long with a temperature sensor after the quencher can be used. The sensor signals are monitored and the rate of change of both sensors is used to trigger the supplemental water addition. In other words, if the sensor closest to the hot source rises rapidly, an output signal is sent to the larger solenoid valve. The sensor after the quencher also tracks the temperature. If the post-quencher temperature continues to rise an additional solenoid valve can be opened. Control valves are often avoided since their response time is slower and they are more expensive. In addition, these solenoid valves may also be used to administer gravity fed water in case of an emergency. To reduce water hammer, pilot operated solenoid valves are sometimes selected. The emergency valve is typically normally open so that under power loss water can flow for cooling purposes.

5. Thermal expansion, particularly in alloy quenchers, can reduce the life of this often expensive piece of hardware. Alloy is expensive and the quencher vessel usually uses as little of the alloy (thickness) as possible. This means that the vessel is an essentially a piece of ductwork designed to support itself with no external load. It is not a structural member. Expansion joints at the gas inlet are therefore commonly used if the design calculations reveal excessive expansion is possible and the vessel is supported in structural steel. Another technique involves "hanging" the quencher from higher-level structural steel and allowing the quencher to "grow" downward. Often an expensive expansion joint at the gas inlet can be avoided. However,

if the quencher is attached to downstream equipment an expansion joint may be needed at the connection of the quencher outlet and that downstream equipment. Hanging the vessel can also make header access easier thus simplifying maintenance.

6. Local corrosion issues usually occur where liquid can "pool" and thereby create a corrosive galvanic cell. When doing an interior inspection, care should be taken to discover and repair any areas wherein liquid can pool. Horizontal surfaces in a vertically oriented quencher should be avoided if possible. If corrosion is noticed on the outside of the quencher, an inspection (if accessible) on the related interior surface should be made. Instead of patching from the outside, patching from the inside may produce a better result. With refractory lined quenchers, a multiple layer type construction may be used. That construction may include an elastomer lining (often rubber) between the vessel wall and the refractory. Sometimes rammed or cast refractory is used. Inspection should include finding and repairing cracks that could allow corrosive water to pass behind the refractory. Though in operation the vessel wall temperature may be within guidelines, the corrosive water could be gradually reducing the quencher's life. Keeping the quencher vessel in optimum condition helps to optimize its performance.

31

Spray Tower Optimization

Spray towers basically disperse by using pressurized spray a high surface area liquid stream counterflow (primarily) into a gas stream containing contaminants and/or heat. The purpose of the design is to efficiently affect the transfer of mass and/or heat into the liquid. The liquid is then constantly drained from the spray tower. The word "primarily" is used because some of the liquid is inherently carried upward in the vessel (sometimes called "back-mixing") given the movement of the gases. Unless the contaminant is absorbed quickly and is rapidly neutralized (retained in solution by being converted to a low partial pressure salt or oxide), movement of the liquid upward can reduce the net efficiency of the spray tower. The optimization goal is therefore to minimize the amount of liquid moving up with the gas stream.

The spray tower shown in Figure 31.1 (Bionomic Industries) was installed on a power boiler.

31.1 Possible Problems

Some possible problems are given below:

1. Excessive vessel vertical velocity thus aggravating back-mixing
2. Improper liquid distribution
3. Single zone application of chemicals, if used, is insufficient
4. Local spray nozzle wear
5. Poor droplet control.

31.2 Optimization Techniques

Some optimization techniques to consider are given below:

1. Ensure the proper vertical velocity of the gases in the tower to minimize back-mixing

FIGURE 31.1
Spray Tower, Bionomic Industries.

2. Provide a uniform dispersion of the liquid across the entire vessel
3. Use "staging," if needed
4. Maintain the integrity of any spray nozzles
5. Control any droplet carryover that could reduce the tested performance.

31.3 Vertical Gas Velocity

Regarding the vertical gas velocity, spray towers can operate as low as about 4–5 ft/s vertical velocity (based upon the actual gas volume being moved) up to 10–15 ft/s. In general, if the designer uses high-pressure spray nozzles that generate exceptionally fine droplets, the vessel velocity is often kept in the low rate to minimizing back-mixing. Conversely, if large droplets were applied using lower pressure nozzles, the vertical velocity chosen would be in the higher range. Obviously, much of the cost of the scrubber is in the vessel thus the designer needs to balance the requirement of adequate performance versus the cost. Spray towers thus tend to be designed in the higher gas velocity range. What can occur is that to achieve greater performance, smaller droplet nozzles are used. What can occur is that doing so actually reduces the performance since the high-pressure nozzles produce smaller droplets that are carried upward. Often a better optimization approach is to increase the liquid to gas ratio (more liquid) but while using nozzles that do not generate fine residual spray. Contacting the nozzle supplier may allow a switch to a different nozzle design (though a pump change may be needed). One must keep the vertical gas velocity within the range as required by the designer. It is usually better to "tweak" the nozzle selection.

31.4 Spray Nozzles and Location

The amount of the spray and the spray's uniform location in the tower are the key to success with a spray tower. The designer typically makes the nozzle location graphically and in doing so there are inherent overlaps. Thus, on a "per unit volume" basis, there will be more sprayed liquid in some areas than in others. Compensation is often made by using multiple spray zones (levels) separated by a distance required to disengage the spray per stage. If the nozzle patterns merge (stage to stage) one can aggravate the situation wherein an excess of liquid occurs in some areas of the tower. To optimize, the spray headers may need to be adjusted to minimize overlap. Some firms provide CFD analysis of such towers. Nozzle vendors may also be able to supply possible location of sprays. In addition, nozzles can often be changed to a different spray angle (rather than changing the header) to alter the coverage. Some stages may use different spray angle nozzles. For example, the upper most spray header may use 120-degree pattern nozzles whereas lower levels may use 90-degree nozzles. As the sprayed liquid descends in the tower, the effect of the nozzle pattern typically becomes less important, but the patterns should be investigated.

31.5 When Chemicals Are Used

In applications that use a reactive chemical to neutralize the absorbed gases, many systems simply administer the chemical to the main recycle header. Though that may be enough for many applications, often the performance can be optimized by also injecting the chemical into a lower header. The latter may involve some relatively minor piping changes. What happens is that the pH in the tower can be better controlled. Fewer products of reaction will drain from the upper spray zone and the driving force will likely be improved at the lower spray zones. A disadvantage in a tower equipped with only a few spray zones could be an increase in unreacted chemical in the scrubber sump. What is usually done is that the gross amount of chemical is kept the same. If the operation reveals greater unreacted chemical in the sump, the amount of chemical going to the lower grid is trimmed back until the system is in balance.

31.6 Spray Nozzle Wear

Spray nozzle wear is a common problem with spray towers but is controlled through proper monitoring of the nozzle performance. If the spray header(s) can be monitored for both pressure and flow, one can interpret from the data the condition of the nozzles. A reduction in pressure and increase in flow usually reveal excessive nozzle wear. An increase in header pressure with a decrease in flow usually indicates plugging. If the pressure and flow is data logged, the trend can be a resource for maintenance personnel regarding the nozzles.

31.7 Back-Mixing Issues

As mentioned earlier, residual spray upward (back-mixing) can reduce the performance of a spray tower. Some back-mixing is likely to occur, but one must not permit those droplets to exit the tower otherwise, during testing, the solids those droplets contain may be counted as an emission, even though the solids are in a droplet. Thus, the droplet control stage (usually a chevron type) is particularly important, Single stage droplet control chevrons may be optimized by converting to two stages. An "interface" tray such as a weeping sieve tray could possibly be added below the main chevron to help agglomerate the spray (make the droplets larger and thus more easily controlled) or a more open, baffle type chevron could be used below the primary chevron to reduce the gross loading of droplets to the primary chevron. If excessive solids are building up on the chevron, a clean in place (CI)P header configuration could possibly be applied to the tower. If used the CIP header usually receives the cleanest water in the system. If scale is experienced, sometimes the CIP circuit is doped with an acid or other descaling chemical.

32

Tray Scrubber (Tray Tower) Optimization

Tray-type scrubbers have been used extensively for gas absorption, gas cooling, and particulate removal. The designs go back to the early 20th century and their characteristics have been well explored over the years. That is not to say that their operation cannot be optimized, however.

32.1 Possible Problems

Some problems that may occur are given below:

1. Inconsistent operation
2. Unusual wear or solids build-up on the trays
3. Weeping (when weeping is not desired)
4. Sudden change in pressure drop
5. Decrease in efficiency (particularly if a change in the application has occurred)
6. Solids build up on the underside of the lowest tray
7. Poor droplet separation.

32.2 Optimization Possibilities

Some things to investigate given the need to improve the operation are the following:

32.2.1 Mounting the Vessel True Vertically

There is a particularly good reason that the scrubber vendor requires that the scrubber be mounted vertically. By "vertically" they mean as measured at the tray level (not, necessarily at the vessel). Inside the scrubber, the scrubbing liquid is flowing across the tray. That liquid flow is influenced by gravity and must be uniform if the tray is to operate properly. If the tray is not level, inconsistent operation can occur. The liquid depth above certain portions of the tray may be lower than at other portions and therefore the gas flow will take the path of least resistance through that lower resistance zone. Higher gas velocities can reduce the mass transfer at that portion of the tray. A tray that is not level could experience a loss in efficiency given poor liquid distribution. Thus, it is imperative that the trays be level. The vessel, however, may not be "true" vertical, thus the tray "decks" should be checked internally to ensure that the trays are level. The remedy may include shimming the tray or even the entire vessel. Another "remedy'" may include adding an "end weir" on the tray (at the point where the liquid exists the tray) to basically "dam up" some of the liquid to force a uniform distribution of the liquid. One should check with the tray vendor and/or consultant, however, since adding liquid depth will increase the pressure drop across that tray. Usually, any end weir is only about 1" (2.5 cm) tall or less. The pressure drop across a tray is a function of the liquid depth and the frictional losses through the openings (usually perforated holes) through the tray. Thus, increasing the liquid depth can directly increase the tray pressure drop. The tray vendor usually applies empirical correction factors to calculate the overall tray pressure drop. It is therefore best to check with the tray supplier prior to adding any end weirs.

32.2.2 Localized Wear and/or Solids Buildup

If upon inspection, unusual wear and/or solids build-up is noticed, it is suggested that those areas be documented (photographed or otherwise noted). Often, localized wear is a function of poor gas distribution that may be a function of the operation of the tray below the problem tray. In other words, once a stream of higher velocity gas is established within the scrubber, there may not be enough "natural" correction forces available in the scrubber to make a correction. The tray scrubber design usually avoids the use of baffles, etc., to provide specific gas flow patterns. The assumption at the design stage was likely uniform "plug" gas flow. The gas pattern was assumed to be uniform across the face of the tray (or trays). High local wear or solids build-up usually indicates a poor distribution of gases. Corrective methods often include investigating the liquid distribution on the "problem" tray and the tray (if any) above that problem tray. As in "a" above, the tray is usually checked to make certain the tray is level and, if level, a study is made as to how the liquid is distributed across the tray. Sometimes solids can build up in the liquid entrance weir (where the recycle liquid enters and is distributed) thus causing poor liquid distribution from the start. Cleaning that liquid entrance weir box area can often solve the problem. If wear is noted at the tray perforations, excessive "normal" gas velocity may be the cause. The tray supplier could possibly recommend trays with large perforations but would need to tune the tray selection to the limits of pressure drop. If the application inherently shows excessive wear, a pre-scrubber such as a venturi type may be needed. That device would reduce the gas inlet particulate loading to reduce the amount of particulate entering the tray scrubber. It is not unusually to see venturi scrubbers being installed as a "team-mate" for a tray scrubber. Another wear producer is operating the tray scrubber at excessive recycle solids content. Usually, a limit of about 1% suspended solids (by weight) is applied to a tray scrubber recycle liquid circuit.

32.2.3 Weeping

Tray scrubbers operate in a range between "weeping" (wherein the gas velocity is lower than that required to keep the liquid above the tray) to "flooding" wherein the gas velocity is so high that the liquid is ejected "overhead" (either upward to the next tray or upward to the droplet separation stage). Weeping can occur (unless desired with a weeping sieve tray) if the gas velocity (in effect kinetic energy) falls below the velocity through the perforations required to hold up the liquid. Obviously, the size of the perforations and the gas density greatly affect the required kinetic energy. If a tray weeps, then perhaps smaller diameter perforations may be required, or blank offs can be applied to the lower surface of the tray. The tray supplier could possibly recommend and supply replacement trays that resist weeping. If blank offs are contemplated, one must be certain that the installation of the blank offs does not adversely affect the gas flow pattern. A blank off applied to a lower tray in a multiple tray scrubber may upset the gas flow through not only the tray in question but also the ones above that tray. Usually, however, the amount of blanking is minor (often less than 10% of the tray) to increase the gas velocity through the perforations of the remaining open portions of the tray to reduce or eliminate weeping.

32.2.4 Sudden Pressure Drop Changes

With a tray scrubber, a sudden reduction in pressure drop usually occurs wherein the "downcomer" seal located ahead of a tray is no longer able to seal. The overall scrubber pressure drop will decrease as the resistance to flow across a tray is reduced. The downcomer seal is in effect a liquid trap (much like an under-sink trap) that normally prevents gases from sneaking up the downcomer without passing through the tray. If the application gas density or gas flow rate increases, the seal height required may be insufficient. Often, adding some height to the seal trap can recover the required seal depth. Also, under scheduled maintenance, the downcomer area(s) should be cleaned (but are often overlooked since they can be difficult to inspect). That seal area can usually be flushed, or power washed without complete inspection. Another cause of a reduction in pressure drop can be the corrosive failure of a tray. If a hole develops in a tray, both the resistance to flow caused by the gas and that of the liquid occurs at the same time. The hole need not be exceptionally large to cause a significant pressure drop reduction. Obviously if an inspection reveals a damaged tray then that tray should be replaced. Yet another cause of a change in pressure drop could be the result of a flooding incident in the scrubber. The trays are typically either bolted in place or wedges are used to secure the tray. If flooding (even for a short period of time) occurs, the forces imparted by the surging liquid can be quite violent causing a tray (or trays to become dislodged. One moved, the tray may not return to its original and proper) location. If the tray is not damaged, the tray can often be returned to its proper position. If the tray is no longer flat, the tray would need to be returned to a flat plain or be replaced. A sudden increase in pressure drop usually is a result of a "puff" of solids from the process blinding the lower surface of the tray. Once the perforations are plugged, many tray scrubbers (unless they are equipped with lower tray "face" sprays that spray upward to clean that tray surface) lack any mechanism to reduce plugging. Given a drawing of the scrubber, it may be possible to add face sprays to an existing tray scrubber. The water balance would need to be checked however to avoid flooding of the lower tray since some of the face spray liquid will go upward through the tray perforations and add to the inventory of the liquid above that tray and perhaps increase that tray's pressure drop. Excessive solids in the recycle liquid can also increase the pressure drop.

32.2.5 Sudden Changes in Efficiency

A reduction in efficiency (absorption, cooling, particulate removal) could be caused by any (or even all) of the above. The preceding are "mechanical" type effects whereas the overall performance of a tray scrubber may be a combination of both mechanical and chemical actions. If the scrubber is used for absorption followed by neutralization or oxidation of absorbed contaminants in the liquid phase, adjustments to the chemistry of the recycle stream may be required. These adjustments are like those that can be applied to packed towers therefore the reader is directed to that section. Another thing that can occur with a tray scrubber that can reduce the efficiency is when the gas velocity is increased to a point wherein the liquid on the tray goes "overhead". Instead of the liquid going across the tray and down (counter-flow) the liquid can be carried upward (to the next tray above or even out to the droplet eliminator). The tray may not be flooding however the distance between trays is such that the liquid cannot separate and fall back but rather be carried up into the upper tray. When that occurs, the liquid above is now contaminated with the liquid from below (dissolved solids, temperature, etc.) thus stage wise separation is reduced. The tray spacing however is usually a "given" and is an inherent part of the vessel tray spacing, thus the optimization techniques are limited. Perhaps larger tray perforations could be used (to reduce the kinetic energy of the gas thereby reducing the height to which the liquid can rise) or the amount of gases being delivered to the scrubber can be reduced (such as precooling and condensing the gases ahead to that scrubber).

32.2.6 Face Spray

If the tray scrubber is applied to a process that emits particulate, the scrubber likely is equipped with a "face spray" as mentioned above. The face spray area usually features full cone spray nozzles whose pattern is selected to completely wash the lower portion "face" of that tray. If possible, that spray circuit should use clean water but too often it does not. If build up is experienced, sometimes the spray circuit solids content can be "cut" by blending in fresh (make up) water to that header circuit rather than adding make up water to the scrubber sump. The fresh water would therefore be put to beneficial use prior to draining to the sump where it would end up anyway.

32.2.7 Droplet Eliminator Effects

Ineffective draining of the droplet control device can cause poor droplet separation. Tray scrubbers often are equipped with "vane type" droplet eliminators. These look like large turbine blade assemblies. These devices use near radially oriented vanes (the vanes are installed tangent to a central cylindrical core) that impart a centrifugal action to the rising gas stream. That action forces the entrained droplets towards the vessel wall from which point the droplets drain through traps to the tray below (or sometimes out of the vessel). If those traps become plugged, the peripheral zone near the vessel wall becomes flooded and re-entrainment of liquid can occur. To optimize the droplet control, these traps must be kept free flowing. Some scrubbers use a timed spray of fresh water to periodically spray the vanes and thus momentarily flush the drains as a purge. This is essentially a clean in place type procedure that supplements periodic maintenance. Of course, if the vanes become damaged that droplet separation could be reduced. If the droplet separation cannot be easily improved through vessel repair and/or cleaning, then a supplemental cross flow droplet separator could be a possible remedy (see other sections where this technique is described).

33

Venturi Scrubber Optimization

The moniker "workhorse" certainly applies to the venturi-type wet scrubber for air pollution control. It got its reputation (and application in thousands of wet scrubber applications) given its simplicity, reliability, and low-capital cost. Like its packed tower brethren, its simplicity makes it sometimes difficult to optimize. Optimization can result in some added cost that is therefore in conflict with its reputation for low cost. Also, as mentioned in the chapter on Venturi scrubbers, improving its removal efficiency often results in an increase in power consumption whether that increase occurs at the prime mover (usually a fan) or at the pump.

There are some things to be considered however to optimize and integrate a venturi scrubber with a process. Usually the first thing to do is to check the pressure (or vacuum) capability of the prime mover (fan, educator, etc.) since optimizing the performance of a venturi type scrubber may require some additional energy input (i.e. pressure drop). The goal, however, is to get the best performance at the lowest energy consumption.

A multiple throat venturi and separator is shown in Figure 33.1 in a typical configuration wherein the venturi section (in this case multiple throats) is in one section and the droplet control is in another. In this case, the droplet control section uses chevron type control thus the containing vessel is enlarged to reduce the "face" velocity of the droplet-laden gas stream.

33.1 Possible Problems

Some typical problem areas may be as follows:

1. Less than desired fine particulate removal
2. Insufficient turndown with respect to gas volume
3. Unstable draft control (noticeable on draft sensitive applications)
4. Plugging in the liquid circuit (particularly associated with designs that use spray nozzles)
5. Wear
6. Poor absorption of soluble gases

FIGURE 33.1
Venturi scrubber, BACT.

33.2 Possible Remedies

Possible remedies to consider include the following:

33.2.1 Preconditioning the Gas Stream

For improving fine particulate removal, "preconditioning" the particulate prior to the venturi throat may help. Preconditioning essentially involves growing the particles so that, given their increased size, they are easier to remove. The "growth" may sometimes be achieved using an air atomized clean water mist or fog as far ahead of the venturi throat as possible. The distance is needed because the particulate merges with the fog in a random fashion so extra time is needed to contact the fog with the particulate. If the liquid added prior to spraying can be converted to an electrolytic solution (for example by adding a halide), the mist or fog can possibly be administered with a static charge (electrostatic spray) that can encourage the fine particulate to combine (agglomerate). A water fog followed by rapid cooling in the venturi (say, using colder recycled liquid) can possibly harness ther-mophoretic forces that tend to move particulate and condensate towards a colder surface. Though these forces are weak, such application may be worth exploring particularly if the gas stream exhibits high humidity (large amounts of water vapor to be utilized). Another possibility is to saturate, and sub cool the gases to induce the water vapor to condense on the fine particulate thus making the particle aerodynamically larger (the so-called "flux force condensation" technique described elsewhere in this book and other literature). Though usually of modest positive effect, surfactants can sometimes be used to "wet" the particulate. The surfactant can be adminis-tered by air or hydraulically atomized sprays applied well in advance of the venturi throat. Another possibility, though not common, is that the venturi throat length is insufficient. What can happen is that the liquid in the ven-turi throat does not reach a point wherein the maximum density of droplets is not achieved. The gases are moving so fast that a "blow hole" is formed in the throat area thus providing a highway through which particulate may pass and escape capture. That zone is usually located at the point where the discharge of the throat zone expands, and the net gas velocity is reduced (sometimes called the "velocity pressure recovery" zone). A possible opti-mization method is to install a spool piece at the discharge portion of the venturi throat that increases the effective throat length so that the "blow hole" closes within the throat thus providing the needed high droplet den-sity zone. It is beyond the scope of this book to explain the intricacies of the throat length design however the original scrubber vendor and/or con-sulting engineering firm may be able to optimize the venturi scrubber if the throat length is suspect. If a rectangular, fixed venturi throat exceeds about 8" width and the liquid enters using a "dentist bowl" type (swirling liquid path) approach to the venturi, sometimes simply installing a vertical plate in the center of the throat can improve performance. What happens is the throat is in effect converted to two parallel narrower throats. The swirling liquid is forced to move through the narrower passageway and thus is distributed more uniformly. Also, the wetted surface of the throat is increased thus providing a slower moving film of liquid on the plate to enhance interception of particulate. Given the greater surface, the pressure drop will increase slightly but that is often compensated by a reduction in the gross liquid rate.

33.2.2 Turndown and Pressure Drop Maintenance

For turndown with respect to gas volume, there are a variety of techniques to explore. One common method is to use an adjustable venturi throat if the venturi scrubber was not so configured initially. Usually the original scrubber vendor (or consulting firm) can recommend and supply an adjust-able venturi throat that can replace the fixed throat. The adjustable throat could be "manual" in that one must move a lever (either by hand or using an positioner) to change the throat pressure drop or "automatic" wherein a signal such as the pressure drop across the throat is sent to an electric, hydraulic or pneumatic positioner to adjust the pressure drop. Another method involves sending some previously scrubbed gases back to the venturi scrubber inlet. The gas flow is modulated using an opposed blade damper equipped with a positioner. The control signal is usually the pressure drop across the venturi throat (often defined in the permit to operate). If control over a fine range (say less than 10% peak gas volume to minimum gas volume) is required, the liquid recycle rate can sometimes provide the needed control. The liquid passing through the throat zone influences the pressure drop across the throat. The greater the liquid rate, the higher the pressure drop. If a draft signal is available (and the Permit to Operate allows it), the liquid flow rate can be used to fine tune the scrubber pressure drop and process draft. The draft signal is used to actuate a control valve that varies the liquid rate (within permitted bounds) to provide the control. The scrubber vendor should be contacted first however to determine the maximum and minimum recycle rates required to meet their performance guarantee. If a modulating damper and a variation in the liquid rate is used for extreme fine-tuning, the damper is used for "gross" control and the liquid rate is used for "fine" control. The control algorithm is configured to modulate the damper less frequently than the liquid rate. Very precise draft or pressure drop control can be obtained. If the facility building is heated or cooled, and the source is located inside the building, precise draft control indirectly minimizes ambient air infiltration that, perhaps at a cost, needs to be heated or cooled. Some operating costs can potentially be obtained.

Somewhat related to the above, for draft sensitive application like hazardous waste incinerators, it often is best to leave the venturi scrubber "fixed" and just finely control the gas volume. The venturi scrubber hydraulic characteristics usually do not respond linearly to adjustments. Adjustable surfaces rotate or swing in arcs yielding a somewhat sinusoidal response instead. If the draft on the source is monitored, however, cleaned gases can often be recycled back to the scrubber inlet based upon that draft signal. The scrubber in effect "sees" a constant gas volume as part of that recycle loop even though the inlet and outlet gas volume may change. Very precise draft control can be achieved if an opposed blade damper is used to regulate the loop volume. Opposed blade dampers provide a more linear response to control versus parallel blade dampers. To reduce the unnecessary "hunting" motion of the damper blades (excessive modulation), the output signal of the draft or pressure drop transmitter is often dampened by configuring the controller to send output signals to the damper positioner less frequently. Less frequent output signals can also accommodate to some extent the compressibility of the gas stream thus smoothing the gas flow through the system. Some installations also use a variable frequency drive on the prime mover to set the "coarse" volume adjustment and use the opposed blade damper configuration to provide the required "fine" adjustment. There are dozens of installations using this technique. Adjusting the liquid rate (mentioned previously) can provide even finer adjustment for extremely draft sensitive applications.

33.2.3 Method of Liquid Introduction

Some venturi scrubber designs use open pipes to deliver the scrubbing liquid, some use spray nozzles and some use a mixture of both. If the scrubbing liquid is essentially "clear" (say just a fraction of a percent solids) then quite often spray nozzles are reliable and plugging resistant. If the solids level is higher either in normal operation or in upset conditions, nozzles can plug. If chronic nozzle plugging occurs the various nozzle suppliers can be consulted to see if a more plugging resistant design can be applied. If the venturi water circuit can be isolated, various solids reduction techniques may be able to be applied to that circuit. Remedies such as increased settling capacity, liquid cyclones; polymer addition, etc. could be candidates for enhanced particle removal. If available and economical, sometimes merely increasing the blow down rate combined with adding any fresh liquid immediately in advance of the spray nozzles can help reduce plugging and serve to optimize that circuit. If the plugging is severe, it may be possible to replace a spray-nozzle based venturi design with one that uses open pipes. To do that either drawings or accurate field measurements are needed to allow the fabrication of a replacement. Often the existing recycle pump is more than adequate (particularly with respect to pressure) since an open pipe liquid circuit usually operates at only a fraction of a spray nozzle type. If the venturi design is already an open pipe type yet plugging occurs, a subtle cause may be solids build up in the feed header (or headers) used to administer the liquid. Some headers feed from the bottom that can lead to a poor distribution of solids. By feeding the liquid to above the injection point and requiring the liquid flow to be down into the scrubber tends to flush solids continually into the scrubber. Doing so often reduces local solids build up that could lead to plugging. If plugging is chronic (such as with processes that inherently must deal with high solids loadings), installing access ports and/or removable pipe sections may be the only way to optimize the liquid circuit. Some facilities use flexible hoses in the liquid recycle circuit to reduce plugging. The hoses are flexed periodically to dislodge built-up solids.

33.2.4 Effects of Wear

Wear is usually a result of a combination of erosion and corrosion in venturi type scrubbers. The wet environment inside the device coupled with the suspended solids that are often present conspire to cause wear particularly in any high gas velocity or direction change zones. Venturi scrubbers often operate at high gas velocities, so a tradeoff occurs between particulate removal efficiency and wear. Efficiency usually controls that debate however the wear becomes an operational "given". As part of the design wear plates (higher hardness metal, elastomeric linings, or refractory type material) can sometimes be applied. On a practical basis, however, usually maintenance-scheduling measurements are taken, and repair panels or components are replaced as part of scheduled maintenance procedures if wear is chronic. To optimize rather than degrade the performance, however, the repairs should be made to duplicate the original interior surfaces as much as possible. Poor welding or abrupt interior surfaces will quickly wear perhaps faster than the original. Regarding corrosion, if chronic corrosion occurs then a change in material of construction (MOC) may ultimately be required. Test coupons could be installed in the existing scrubber to best determine the replacement MOC rather than guessing. Published corrosion resistance data may be helpful but not for the specific process at hand. Rarely is the corrosive environment exactly as described in corrosion resistance charts or literature. Sometimes the liquid is aerated or otherwise modified during contact in the scrubber that changes the overall corrosion characteristics. For example, an infeed neutralization chemical may eventually become a corrosive salt in the scrubber. If the products of reaction are not sufficiently bled away, local concentrations can exceed those described in the MOC corrosion resistance literature. For those reasons and others, wet scrubber vendors do not typically offer a guarantee against corrosion. To optimize the MOC to suit the application local testing with coupons or bench scale evaluation is recommended rather than a literature search.

33.2.5 Improving Gas Absorption

Venturi scrubbers provide modest gas absorption along with superior particulate removal. The modest absorption is a result of the concurrent motion of the liquid and gas coupled with the relatively short contact time. The gas stream in a venturi scrubber may be moving at over five to ten times the speed of the gas in a packed tower. In a packed or tray tower, the gas and liquid pass counterflow so that the cleanest gas "sees" the cleanest liquid. In a venturi scrubber, the gas and liquid share the same highway in the same direction. Unless the contaminant gas is reduced in partial pressure through chemical conversion to a low vapor pressure compound, the gas can leave the highway without being absorbed. Given those inherent limitations, however, gas absorption in venturi scrubbers can be optimized by controlling the recycle liquid pH and temperature much as described in the section on packed towers. In addition, sprays situated well in advance of the venturi scrubber can provide additional mass transfer surface for gas absorption. Fresh chemical added into that spray circuit rather than into the sump recycle circuit could add to the gas absorption capability of the venturi scrubber. If the scrubber uses a cyclonic type separator for droplet control, a spray header can sometimes be added at the inlet to the separator to provide additional liquid to gas surface area for mass transfer of the gas into the liquid. The spray droplet size is chosen to be the same or larger than the droplets that are expected to be produced by the venturi. If smaller droplets are formed by the spray(s) an additional droplet separation is placed on the cyclonic separator. Also, the drain in the separator needs to be checked to make certain that the increased liquid rate provided by the spray does not overwhelm the drain. If the drain capacity is marginal, the drain can be enlarged, or a supplemental drain can be added. The drain line size external to the scrubber must be adequately sized. Typically, an additional spray volume of 5%–10% of the initial recycle volume can be accommodated by an original drain size (given that drains are usually oversized). If the added spray is predicted to be greater, a thorough analysis of the drain piping should be entertained.

Appendix A: Additional Selected Reading

The following is a list of books and publications that are often seen on the shelves of professional air pollution control personnel. For more detailed information about a product, application, or gas cleaning technique, these references will be of great value to you.

A listing of the details of the individual publications is at the end of this appendix.

GENERAL TOPICS

INDUSTRIAL VENTILATION: A MANUAL OF RECOMMENDED PRACTICE

This classic work is a valuable reference regarding gas collection and movement techniques. In print since 1951, it contains information regarding collection hood sizing, conveying velocities, ductwork friction losses, contaminant exposure limits, and the ventilation aspects of industrial hygiene.

Air Pollution Engineering Manual

As an update of the old AP-42 U.S. government publication regarding the application of air pollution control devices, this essential resource contains a detailed compendium of application descriptions by industry written by a variety of experienced designers and application engineers. A wealth of practical and useful information is contained therein.

Fan Engineering

Produced by Howden Fan Company, this power-packed book contains excellent information regarding gas flow rates, gas-moving devices (such as fans), air pollution control hardware, psychometrics, and related air/gas properties.

Industrial Research Service's Psychometric Tables and Charts

If you are not familiar with the properties of air and the moisture it can carry, this book by Zimmerman and Lavine (1953) might appear a bit daunting. Though computerized gas mixture property predicting programs are now available (see *Psychrometric Problem-Solving Program* under the "Publication Details" section that follows), the Psychrometric Tables and Charts are still in daily use by air pollution control professionals. With these charts and tables, one can accurately predict gas mixture properties that form the basis of gas cleaning system design.

The McIlvaine Scrubber Manual

This is a comprehensive manual in the form of multiple binders plus a newsletter all available on a subscription basis. It is excellent for people or firms who deal with air pollution control problems repeatedly during the year. It is of great value as well for people who must keep up to speed with the latest advances in pollution control. Highly recommended.

Cameron Hydraulic Data

Particularly useful regarding wet scrubbers, this classic reference provides excellent information regarding pumping, piping, and frictional losses.

Mass Transfer Operations

Few books on mass transfer are as widely used as this famous book by Robert E. Treybal. Often used as a textbook, it is found on the shelves of pollution control professionals or process designers whose job is to design equipment that moves a gas (or heat) into or out of a liquid.

Various Corrosion Guides

Too numerous to mention specifically by name, several pump and/or piping material suppliers publish corrosion guides for the application of their products. These are guides, however, and do not offer guarantees of material of construction applicability. The suggested thing to do is to accumulate a variety of them and look for a consensus as to materials deemed suitable for the application. A few of the more popular guides are listed in the following section.

PUBLICATION DETAILS

The following is a list of publication details for the items mentioned above plus a few other periodicals and resources you may consider for your library.

Air Pollution Control

Traditional and Hazardous Pollutants

Dr. Howard E. Hesketh

Technomic Publishing Co.

CRC Press

6000 Broken Sound Parkway NW, Suite 300

Boca Raton, FL 33487

Air Pollution Engineering Manual

Anthony J. Buonicore and Wayne T. Davis, editors

Van Nostrand Reinhold Publishers

115 Fifth Avenue

New York, NY 10003

Atlac Guide to Corrosion Control

Reichold Chemical Company

P. O. Box 19129

Jacksonville, FL 32245

Bete Fog Nozzle Catalog

50 Greenfield Street

Greenfield, MA 01302-0311

Cameron Hydraulic Data

C.R. Westaway and A.W. Loomis

Ingersoll Rand

Woodcliff Lake, NJ 07675

Chemical Engineering

Chemical Week Publishing

P. O. Box 619

Mt. Morris, IL 61054-7580

http://www.echm@kable.com

Derakane Chemical Resistance Table

Dow Chemical Company

2040 Willard H. Dow Center

Midland, MI 48640

Dwyer Instruments, Inc.

102 Indiana Hwy.

P. O. Box 373

Michigan City, IN 46360

http://www.dwyer-inst.com

219-879-8000

Fan Engineering

Howden North America, Inc., Corporate Office

7909 Parklane Road

Columbia, SC 29223

1-803-713-2200

Sales@howdenbuffalo.com

Guide to Corrosion Resistance, A

Climax Molybdenum Company

One Greenwich Plaza

Greenwich, CT 06830

Handbook of Separation Techniques for Chemical Engineers

Phillip A. Schweitzer, editor

McGraw-Hill

1221 Avenue of the Americas

New York, NY 10020

Huntington Alloys Corrosion Chart (Nickel Alloys)

Ask local representative or write to:

Huntington Alloys Corp.

3200 Riverside Drive

Huntington, WV 25705

Industrial Research Service's Psychrometric Tables and Charts (1953)

O. T. Zimmerman and Dr. Irvin Lavine

Industrial Research Service, Inc.

Dover, NH

*Industrial Ventilation: A Manual of
    Recommended Practice*
American Conference of Governmental
    Industrial Hygienists
6500 Glenway Avenue
Bldg. D-7
Cincinnati, OH 45211

*Journal of the Air and Waste Management
    Association*
P. O. Box 2861
Pittsburgh, PA 15230

Mass Transfer Operations
Robert E. Treybal
McGraw-Hill
1221 Avenue of the Americas
New York, NY 10020

McIlvaine Scrubber Manual, The
The McIlvaine Company
2970 Maria Avenue
Northbrook, IL 60062

Pollution Engineering Magazine
Cahners Business Information
8773 S. Ridgeline Boulevard
Highlands Ranch, CO 80126
http://subsmail@cahners.com

Power Engineering
PennWell Corp.
1421 S. Sheridan Road
Tulsa, OK 74112
http://www.power-eng.com

*Psychrometric Problem-Solving
    Program*
DesJardins and Associates
38373 Cherrywood Dr.
Murrieta, CA 92562
richard@rjdesjardins.com

Stainless Steel in Gas Scrubbers
Committee of Stainless-Steel
    Producers
American Iron and Steel Institute
1000 16th Street, NW
Washington, D. C. 20036

*Technical Association for the Pulp and
    Paper Industry*
TAPPI Journal
15 Technology Parkway, South
Norcross, GA 30092
http://www.tappi.org

Appendix B: List of Photo and Content Contributors

Air-Clear, LLC
2440 Oldfield Point Road
Elkton, MD 21921-6712
www.airclear.net

Adwest Technologies, Inc., CECO
  Environmental
690 Langsdor Drive, Suite 102
Fullerton (University Plaza), CA 92834

Air Instruments and Measurements,
  Inc. (AIM)
13300 Brooks Dr. Suite A
Baldwin Park, CA 91706
www.aimanalysis.com

Allen-Sherman-Hoff (ASH)
457 Creamery Way
Exton, PA 19341
www.babcock.com

Alzeta Corp.
2343 Calle del Mundo
Santa Clara, CA 95054-1008
www.alzeta.com

American Air Filter
AAF International
Suite 2200
Louisville, KY 40223-5000
www.aafintl.com

Amcec, Inc.
2525 Cabot Drive
Suite 205

Lisle, IL 60532
www.amcec.com

Babcock & Wilcox Company
1200 E. Market Street, Suite 650
Akron, OH 44305
www.babcock.com

BACT Process Systems Inc.
3345 N Arlington Heights Road
Suite B
Arlington Heights, IL 60004
www.bactprocess.com

Banks Engineering, Inc.
3715 East 55th Street
Tulsa, OK 74135
www.banksengineering.com

Barnebey Sutcliffe Corp.
835 Cassady Ave.
Columbus, OH 43216-2526
www.bscarbons.com

BHA Group, Inc.
PrecipTech, Inc.
8800 E. 63rd Street
Kansas City, MO 64133
www.bhagroup.com

Bindicator
150 Venture Blvd.
Spartenburg, SC 29306
www.info@bindicator.com

Bionomic Industries, Inc.
777 Corporate Drive
Mahwah, NJ 07430
www.bionomicind.com

Bremco
P. O. Box 1491
Claremont, NH 03743
www.bremco.com

Brentwood Industries, Inc.
500 Spring Ridge Drive
Reading, PA 19611
www.brentwoodindustries.com

Bundy Environmental Technology,
    Inc.
921 Eastwood Dr., Suite 115
Westerville, OH 43081
www.bundyenvironmental.com

Burt Process Equipment
100 Overlook Dr.
Hamden, CT 06514
www.burtprocess.com

Carbtrol Corp.
200 Benton Street
Stratford, CT 06615
www.carbtrol.com

Claffey
C&W Technical Sales, Inc.
3555 Hillside Road
Slinger, WI 53086
www.cwtech-sales.com

Colannino Associates
Joseph Colannino, CEO
3027 Andorra Way
Oceanside, CA 92056

joecolannino@gmail.com
www.combustion-modeling.com

Donaldson Company, Inc.
Industrial Air Filtration
P. O. Box 1299
Minneapolis, MN 55440-1299
www.donaldson.com

Dwyer Instruments Inc.
102 Indiana Hwy. 212
P. O. Box 373
Michigan City, IN 46360
www.dwyer-inst.com

Entoleter, Inc.
251 Welton Street
Hamden, CT 06517
www.entoleter.com

Envirogen
4100 Quakerbridge Road
Lawrenceville, NJ 08648
www.envirogen.com

Envitech
2924 Emerson Street
Suite 320
San Diego, CA 92106
www.envitech.com

Euro-matic UK Ltd.
71-75 Shelton St.
Covent Garden
London, WC2H 9JQ, U.K.
www.euro-matic.co.uk

Fluid Technologies (Environmental), Ltd.
50 Old London Road
Kingston-upon-Thames, Surrey KT2
    6QF, U.K.
www.fluid-technologies.com

Geoenergy International Corp.
LDX Solutions
8271 154th Ave., NE, Suite 250
Redmond, WA 98052
www.LDXsolutions.com

Hart Environmental, Inc.
4630 Brighton Lake, Dr.
Cumming, GA 30040
www.hartenv.com

John Zink Company, LLC
Div. of Koch-Glitsch, Inc.
11920 E. Apache
Tulsa, OK 74116
www.johnzinkhamworthy.com

Kenics (Chemineer)
5870 Poe Ave.
Dayton, OH 45414
chemineer@nov.com
www.nov.com/mixing

Kimre, Inc.
744 SW 1st St.
Homestead, FL 33030
www.kimre.com

Koch-Glitsch, Inc.
4111 E. 37th Street N.
Wichita, KS 67220
www.koch-glitsch.com

Lantec Products, Inc.
5302 Derry Avenue, Unit G
Agoura Hills, CA 91301
www.lantecp.com

Wm. W. Meyer and Sons, Inc.
Meyer Corporate
1700 Franklin Boulevard

Libertyville, IL 60048-4407
www.wmwmeyer.com

MikroPul, LLC, a Nederman Company
4433 Chesapeake Drive
Charlotte, NC 28216

Misonix, Inc.
1938 New Highway
Farmingdale, NY 11735
www.misonix.com

Monsanto Enviro-Chem Systems, Inc.
   (MECS)
1422 South Outer Forty Road
Chesterfield, MO 63017
www.enviro-chem.com

Munters Corp.
P. O. Box 6428
Fort Myers, FL 33911
www.munters.com

Munters Zeol
79 Monroe Street
Amesbury, MA 01913
www.munterszeol.com

RVT Process Equipment, Inc.
9047 Executive Park Drive, Suite 222
Knoxville, TN 37923
www.rvtpe.com

Sly, Inc.
8300 Dow Cir., Suite 600
Strongville, OH 44136
www.slyinc.com

SRE, Inc.
510 Franklin Avenue
Nutley, NJ 07110
www.srebiotech.com

Steelcraft Corp.
P. O. Box 820748
Memphis, TN 38182-0748
www.steelcraftcorp.com

T-Thermal Company (Selas Fluid
    Processing)
325 Sentry Parkway E
Blue Bell, PA 19422
www.t-thermal.com

TREMA Verfahrenstechnik GmbH
Rohrwiesen 1
D-95478

Kemnath, Germany
www.trema.de

TurboSonic Technologies, Inc.
550 Parkside Drive, Suite A-14
Waterloo, Ontario, Canada
N2L 5V4
www.turbosonic.com

Uzelac Industries, Inc.
Duske Drying Systems
6901 Industrial Loop
Greendale, WI 53129
www.uzelacind.com

Afterword

It is hoped that the foregoing information has helped you select equipment suitable for your application or area of interest and has provided supplemental information and support in the related effort to optimize the operation of the equipment of interest. In addition, may this book reside on your reference shelf in the years to come to support your continuing efforts to control air pollution both for your health and the health of others.

Index